Some Mathematical Questions
in Biology

MODELS IN
POPULATION BIOLOGY

Lectures on Mathematics in the Life Sciences
Volume 20

Some Mathematical Questions
in Biology
MODELS IN
POPULATION BIOLOGY

The American Mathematical Society
Providence, Rhode Island

Proceedings of the 1987 Symposium *Some Mathematical Questions in Biology* held at the Annual Meeting of the American Association for the Advancement of Science in Chicago, Illinois, February 27, 1987. This Symposium was sponsored by the National Science Foundation under Grant No. DMS-8610499.

Edited by

Allan Hastings

1980 *Mathematics Subject Classification* (1985 *Revision*). Primary 92A10, 92A15.

The Library of Congress has catalogued this serial as follows:
Symposium on Some Mathematical Questions in
Biology (1987: Chicago, Ill.)
Some mathematical questions in biology.
(Lectures on mathematics in the life sciences; v. 20)
"Proceedings of the 1987 Symposium [on] Some Mathematical Questions in Biology held at the annual meeting of the American Association for the Advancement of Science in Chicago, Illinois, February 27, 1987. This symposium was sponsored by the National Science Foundation"–T.P. verso.
Includes bibliographies.
1. Population biology–Mathematical models– Congresses. I. Hastings, A. (Alan), 1953- .
II. National Science Foundation (U.S.)   III. Title.   IV. Series.
QH353.S96   1987   574.5′248′011
ISBN 0-8218-1170-3   (alk. paper)

89-15119

CONTENTS

PREFACE

Population biology has had a long history of mathematical modelling, and an early culmination was reached in the 1920's and 1930's with the work of Lotka and Volterra in ecology and Fisher, Haldane, and Wright in genetics. Recently, much more sophisticated mathematical techniques have been brought to bear on the questions of ecology and genetics. Simultaneously, great advances in experimental and field work have produced a wealth of new data.

One of the drawbacks of this explosive growth has been a tendency for the field of population biology to become fragmented. One common feature, however, is that similar mathematical questions and techniques arise in disparate areas of investigation. Further sharing of these mathematical advances among the different areas of population biology will certainly prove useful in the future.

The contributions reported here all deal with different aspects of population biology, but there are overlaps in the mathematical techniques used. Dynamics of nonlinear differential and difference equations form a common theme. Boyd and Richerson consider problems in cultural evolution. Hastings considers problems in multilocus population genetics, and Nagylaki considers problems in spatially structured population genetics. Schaffer discusses questions of chaos and the dynamics of epidemics. Chesson discusses questions concerned with the dynamics of ecological communities.

Alan Hastings

Chicago, Illinois
February 27, 1987

Lectures on Mathematics in the Life Sciences
Volume 20, 1989

SOCIAL LEARNING AS AN ADAPTATION

Robert Boyd and Peter J. Richerson[1]

ABSTRACT. Social learning entails an interesting
adaptive trade-off. Imitating others is useful when
the locally adaptive behavior is common. However, as
a population comes to rely more heavily on social than
on individual learning, the locally adaptive behavior
will become less common. Here, we analyze several
simple mathematical models that embody this trade-off.
We show that a heavy reliance on social learning is
favored when environments are variable but not too
variable. We also show that reliance on social
learning is favored when one behavior is common among
an individual's models.

1. INTRODUCTION. Learning is widespread in the animal kingdom.
While the mechanisms of learning range from relatively simple
conditioning in invertebrates to elaborate cognitive mechanisms in
mammals, most animals use some form of learning to acquire
behavior that is adaptive in the local habitat. Despite this
fact, the great bulk of evolutionary theory assumes that organisms
adapt to variable environments through genetic mechanisms alone.
The neglect of learning may result from the difficulty of
understanding the evolution of learned behaviors. Learning
entails an evolutionary tradeoff. The advantages of learning are
obvious; it allows the same individual to behave appropriately in

---

1. 1980 Mathematics Subject Classification (1985 Revision).

92-A10, 92-A15, 90-A07.

different environments. For example, by sampling novel foods and learning to avoid noxious food types, a cosmopolitan species like the Norway Rat can acquire an appropriate diet in a wide range of environments. However, learning also has disadvantages. First, the learning process itself may be costly. By sampling novel foods, the rat may accidentally poison itself, a risk that could be avoided by an animal with rigid, genetically specified food preferences. Second, because learned behavior is based on imperfect information about the environment, it can lead to errors. For example, the rat may fail to sample or mistakenly reject a nutritious food item. To understand variation in learned behavior among species, one must understand how this evolutionary tradeoff is resolved.

Recently, several authors have used statistical decision theory to understand why the learning rules of different species vary. (McNamara and Houston, 1980, Staddon, 1983, Stephens and Krebs, 1988). One can think of individual organisms as having to "choose" among alternative behaviors so as to maximize their fitness in the local environment. They have some genetically inherited "prior" information about the state of local environment, some data from their experience, and usually the opportunity to gathering more data at some cost in terms of fitness. Decision theory is useful because it tells us the best way to make decisions with imperfect information. Assuming that natural selection has shaped the learning rules of different species so that they are adaptive, decision theory should help us to understand why different animals learn differently. In the same way that mechanics helps us understand the comparative morphology of skeletons, decision theory may help us understand comparative behavior of animals.

We are interested in understanding the adaptive function of one particular form of learning, social learning. By social learning we mean the acquisition of behavior by observation or teaching from other conspecifics. Social learning has been implicated in the acquisition of behavior in a variety of taxa. Many song birds acquire their song by copying the song of other adult birds (Marler and Tamura, 1964). Rats seem to acquire food preferences both from taste cues in their mother's milk, and from the smell of other rat's pelage (Galef, 1976). There is circumstantial evidence that individuals of several different primate species may acquire complex new behaviors by social learning (Kawai, 1965, McGrew and Tutin, 1978, Hauser, 1988). Finally, social learning plays an essential role in human adaptation (Boyd and Richerson, 1985). For reviews of the literature on social learning in non-human animals, see Galef (1976, 1988).

In this paper we present several simple mathematical models of social learning. Our aim is to use these models to understand social learning as an adaptation in the same way that decision theoretic models have been used to understand other forms of learning. The decision theoretic models alone are not sufficient to understand the conditions under which social learning is adaptive. Instead, decision theoretic models must be generalized to allow for the fact that behaviors acquired by social learning are transmitted from individual to individual. Thus to understand social learning, we need models which keep track of the processes that change the frequency of alternative behaviors in a population through time. To see this, consider a young rat learning food preferences. To predict whether it acquires a preference for some food, say cilantro, by social learning, we need to know whether its mother's diet includes cilantro. Its mother's diet will depend on both her experience and her own mother's diet. More generally, to understand why a preference for cilantro among a population of rats is becoming more common (or more rare), we must know its frequency among rats of previous generation, and how this generation's individual learning experiences changed the frequency

of the preference between the time that they acquired their
initial food preferences by social learning and the time that they
serve as models for members of the next generation. Because
behavioral variants are transmitted from individual to individual,
and thus from generation to generation, understanding social
learning requires understanding the dynamic processes that act to
change the frequency of different socially learned behaviors in a
population of organisms through time. We must link models of
individual learning to models of social learning to determine the
evolutionary dynamics of behavioral variants in a population.

We will use these models to address two questions about the
adaptive function of social learning:

A. *Under what circumstances should natural selection favor
increased reliance upon social learning at the expense of
individual learning?* We will begin by analyzing a model in which a
population of organisms acquires behavior by a combination of
individual and social learning in a uniform and constant
environment. This model indicates that, on average, in constant
environments, reliance on social learning always leads to higher
fitness than reliance on individual learning. We will then add
environmental variability to the model. Under these conditions,
there is an optimal mix of social and individual learning. The
relative importance of social learning in the optimal mix is
increased when (1) when environments are predictable, and (2) when
individual learning is error prone.

B. *Given that naive individuals experience the behavior of a
number of experienced individuals, and that this behavior varies,
how should social learning be structured?* Here we will consider a
model in which naive individuals are exposed to a finite sample of
the behavior of members of the previous generation. We will refer
to this set of observed and potentially imitated individuals as
"models." Naive individuals will be exposed to different
combinations of behavior which they can imitate. The analysis

suggests that in a variable environment, selection favors individuals who are predisposed to acquire the most common behavior amongst their models. It also suggests that selection favors individuals whose propensity to rely on individual learning increases as the variability among their set of models increases.

2. A MODEL OF INDIVIDUAL AND SOCIAL LEARNING. We begin by addressing the question: When does social learning allow a more effective tracking of the environment than individual learning? To answer this question, we want to construct a model that embodies the following assumptions about the interaction of social and individual learning:

     A. A population of organisms is potentially confronted with a variable environment in which different behaviors are favored by selection in different habitats.

     B. Individuals in the population can acquire their behavior by some mixture of social learning and individual learning, where:

     C. Social learning involves the faithful copying of the behavior of a single other individual in the population, and:

     D. Individual learning occasionally leads to errors.

     E. All individuals pay any fitness costs associated with individual learning whether they ultimately acquire a behavior by social learning or by individual learning.

Given these assumptions, we want determine the conditions under which selection will favor individuals who rely significantly on social rather than individual learning. Consider a population that occupies an environment that can be in either of two distinct states, labeled habitat 1 and habitat 2. Each individual in the population will acquire one of two alternative behaviors, also labeled 1 and 2. As is shown in table 1, each individual has a "baseline" fitness W; individuals who acquire the behavior that

is best in their environment achieve an increase in fitness, D.
Thus, individuals that acquire behavior 1 have higher fitness in
habitat 1 than individuals that acquire behavior 2. Similarly,
behavior 2 yields higher fitness in habitat 2 than does behavior
1. Once an individual has acquired one of the two behaviors, it
does not change. Nor does the environment change, so that an
individual experiences only one of the two environmental states
during its lifetime.

fitness associated with

	behavior 1	behavior 2
habitat 1	$W + D$	W
habitat 2	W	$W + D$

Table 1

The adaptive problem that faces each individual is to
determine which of the two habitats it is in. Individuals in the
model have two sources of information available to help them solve
this problem.

Each individual obtains evidence from its own experience.
By this we mean any observations, learning trials, or other non-
social information that can help determine the state of the
environment. We assume the result of each individual's experience
can be quantified in terms of a single normally distributed random
variable, x. If the environment is in state 1, the mean value of
x is M; if it is in state 2, the mean value of x is $-M$. In other
words, the true state of the environment is either M or
$-M$. Individuals acquire an imperfect estimate of the state of the
environment, x, from personal experience. The standard deviation
of the distribution of x, S, is an inverse measure of the quality

of the evidence available to the members of the population. The larger is S, the poorer the individual's estimate of the state of the environment. If $S<<|M|$ then most individuals' experiences will clearly indicate the state of the environment. If $S>>|M|$, the results of gathering direct evidence will not be very informative.

Assume that the population is structured into nonoverlapping cohorts. Individuals in one cohort can observe the behavior of individuals from the previous cohort that have already acquired either behavior 1 or behavior 2. Individuals in one cohort act as models for individuals in the next cohort.

We imagine that individuals in the population use these sources of information to decide between the two alternatives in the following way: If the outcome of direct observation, x, is greater than a threshold value d ($d \geq 0$), the individual acquires behavior one; if x is less than $-d$, then it acquires behavior two. This is our attempt to capture the essence of the processes of individual learning. Finally, if $-d \leq x \leq d$, then the individual imitates the behavior of a single individual chosen at random from the population, its model. This, in turn, is our attempt to capture the essence of social learning. The order in which the two kinds of learning occur is not crucial; the model applies equally well to a situation in which individuals begin by imitating others, and then adopt a new behavior only if confronted with decisive personal experience.

The parameter d serves two functions. First, as is shown in figure 1, it is analogous to a confidence interval. The larger the value of d that characterizes the population, the more decisive the evidence must be before it will affect the individual's decision. Second, the value of d simultaneously determines the relative importance of social learning and individual learning. We assume that when individuals are in doubt on the basis of their own experience, they utilize behaviors acquired by imitation. Let p_1 be the probability that $x > d$, and let p_2 be the probability that $x < -d$. If d is large then individuals attend to their own experience only if it provides compelling evidence about the state of the environment (i.e. p_1, $p_2 \approx 0$). For the most part they imitate another individual. If d is small, behavior is mainly determined by an individual's experience, and social learning has little importance (i.e. $p_1 + p_2 \approx 1$).

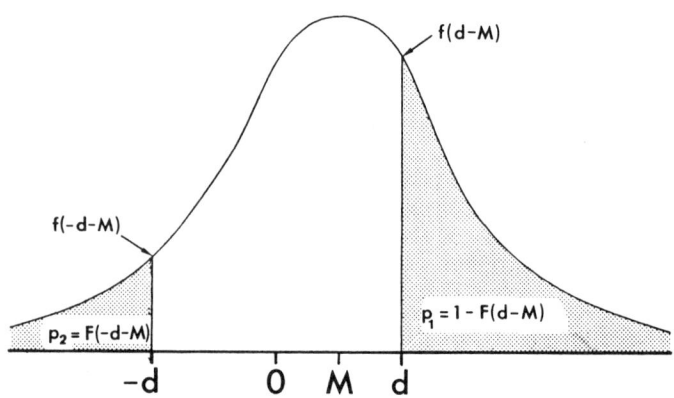

Figure 1: Illustrates the definition of p_1 and p_2, and their relationship to the parameter d. $F(x)$ is the cumulative normal distribution, and $f(x)$ is the normal density function.

2. EFFECTS OF LEARNING ON THE DISTRIBUTION OF BEHAVIOR IN THE POPULATION. To predict the likelihood that an individual will acquire a particular behavior by social learning, we must know what behavior characterizes the individual's model. Suppose that a fraction q_t of individuals in cohort t acquired behavior 1. A

fraction p_1 of the naive individuals in cohort t will acquire
behavior 1 based on their own experience, and a fraction
$q_t(1 - p_1 - p_2)$ acquire alternative 1 by imitation. Thus in
cohort t the frequency of individuals acquiring behavior 1, q_t',
is

(1) $q_t' = q_t(1 - p_1 - p_2) + p_1$

Now suppose that these individuals then serve as models for
individuals in the cohort $t + 1$. Then the frequency of behavior 1
among the models for cohort $t + 1$, q_{t+1}, is approximately:

(2) $q_{t+1} = q_t'$

We say approximately because we have ignored the effect of natural
selection. In environment 1, differential mortality will increase
the frequency of behavior 1. Here we are assuming that the effect
of learning on the relative frequencies of the two behaviors is so
much greater than the effect of selection, that selection can be
safely ignored.

Suppose that this process is repeated many times. That is,
members of a cohort acquire their behavior by a combination of
social and individual learning, then serve as models for the next
cohort, and this process is repeated for many successive cohorts.
Eventually the fraction of each cohort acquiring behavior 1
willstabilize at the equilibrium value

(3) $$\hat{q} = \frac{1}{1 + p_2/p_1}$$

Thus, the fraction of individuals acquiring behavior 1 at
equilibrium depends only on the ratio of the probability that an
individual will choose alternative two based on the its own
experience (p_2) to the probability that it will choose alternative

one based on the its own experience (p_1). If $p_2/p_1 > 1$, then the equilibrium frequency of individuals choosing alternative 1 is less than one half; if $p_2/p_1 < 1$, $\hat{q} > 1/2$. The fraction choosing alternative 1 at equilibrium does not depend (directly) on the relative importance of social learning versus individual learning in determining the behavior of individuals (i.e. on the magnitude of $1 - p_1 - p_2$). However, from equation 1 we know that the rate at which the population converges to the equilibrium value depends crucially on the amount of social learning. If there is little individual learning p_1 and p_2 will be very small and social learning will ensure that the population remains very similar from one generation to the next. Thus as individual learning becomes less important in determining individual behavior, the population will converge more slowly to equilibrium. This property is crucial to understanding the evolution of mixed systems of social and individual learning in variable environments, as we will see below.

3. THE EVOLUTION OF SOCIAL LEARNING. We now consider the evolution of social learning. The relative importance of individual learning and social learning in determining phenotype is given by the parameter d. If d is affected by heritable genetic variation, then it will evolve under the influence of natural selection. We will model the evolution of d using the ESS approach. That is, we assume that an individual's learning rule is affected by a genetic locus at which two alleles, a common allele, H, and a very rare allele, h, are segregating. Most individuals in a population are characterized the genotype HH which result in them having a learning rule characterized by the parameter value d, however, there are a few rare mutant Hh individuals whose learning rule is characterized by a slightly different parameter value, $d + \delta$. We assume that the hh genotype is so rare that it can be neglected. We then determine the

conditions under which the rare allele can invade.   The ESS value
of d is that value which prevents any rare alleles from invading.
When the ESS value of d is very large, we will say that social
learning is adaptive, since when d is large, most individuals will
depend on social learning.

As a first step in understanding the evolution of social
learning, we calculate the ESS value of d, assuming that the
environment is entirely in state 1. In this case, the expected
fitness of an individual whose learning rule is characterized by
the parameter d' in a population in which most individuals have a
learning rule characterized by parameter d (where d' may or may
not equal d) is given by:

(4) $$E\{w(d')\} = W + D\{\hat{q}(d)[1 - p_1(d') - p_2(d')] + p_1(d')\}$$

where $\hat{q}(d)$ is the frequency of behavior 1 at the equilibrium
value given in (3) assuming that most individuals in the
population are characterized by learning parameter d. The rare
allele, h, can invade the population if Hh individuals whose
learning rule is characterized by learning parameter $d + \delta$ have a
higher expected fitness than HH individuals whose learning rule is
characterized by learning parameter d, that is, if
$E\{w(d + \delta)\} > E\{w(d)\}$. Since δ is small,
$E\{w(d + \delta)\} \approx E\{w(d)\} + \delta(\partial E\{w(d)\}/\partial d)$, this condition can be
rewritten in the following form:

(5) $$\delta \left[\frac{\partial p_1}{\partial d} p_2(d) - \frac{\partial p_2}{\partial d} p_1(d) \right] < 0$$

Suppose that the invading allele increases d, so that $\delta > 0$. It
follows from the definitions of d, p_1, and p_2 that a given change
in d causes a larger absolute decrease in p_1 than in p_2, or

$\partial p_1/\partial d < \partial p_2/\partial d < 0$. Thus, inequality (5) says that the rare allele can invade whenever the per cent decrease in the probability of acquiring the wrong behavior by individual learning exceeds the per cent decrease in the probability of getting the right behavior by individual learning. It can be shown that this expression is satisfied for all values of d. This means that the ESS value of d is as large as possible.

We draw two lessons from this simple result. First, some social learning is always better than relying completely on the results of experience. (That is, the expected fitness of an individual using a learning rule characterized by $d = 0$ is always less than the expected fitness of individuals using a learning rule characterized by any positive value of d.) Second, in a population which is characterized by the ESS value of d, individuals may virtually ignore the evidence presented by direct experience and depend entirely on social learning, even when the only cost associated with learning is the occasional error.

It is important to notice that this result was derived assuming that every individual in every cohort experienced habitat 1. This assumption of an invariant environment is crucial because, as we have seen, the equilibrium frequency of the superior variant does not depend on the amount of individual relative to social learning, but the rate of approach to that frequency does. It seems likely that in a variable environment the expected fitness of individuals in the population will depend on the rate at which the population can respond to changes as well as the eventual equilibrium.

4. SOCIAL LEARNING IN VARIABLE ENVIRONMENTS. To introduce
environmental variation into the model, suppose that one-half of
each cohort experiences environment one and the other half of each
cohort experiences state two. (The assumption that the habitats
are the same size greatly simplifies the mathematical argument
without altering the essential aspects of the problem.) Let p_{jk} be
the probability an individual's choice is based on direct
experience and that it results in behavior k given that the state
of the environment is j. Notice that because of the symmetry of
the model

(6)
$$p_{11} = p_{22}$$

$$p_{12} = p_{21}$$

    Variable environments are only interesting in an evolutionary
context if events in one environment affect the other. Migration,
a flow of behavioral variants from one environment into the other,
will likely influence evolution in spatially variable
environments. To model this effect we suppose that there is a
probability $1 - m$ that each model to whom a given individual is
exposed experienced the same environment that the given individual
will experience, and therefore a probability m that the model
experienced the other environmental state. Thus, m measures the
effective rate of migration of individuals from one habitat to the
other. We assume throughout that $0 \leq m \leq 1/2$. Let $q_{t,j}$ be the
fraction of individuals that acquire behavior 1 within the
subpopulation of individuals that experience environmental state j
in cohort t. Then the frequency of behavior 1 in environment j
after learning but before migration will be:

(7) $q_{t,j}' = q_{t,j}(1 - p_{j1} - p_{j2}) + p_{j1}$

and the frequency of models exhibiting alternative t in habitat j during cohort $t + 1$ is

(8)

$$q_{t+1,1} = (1 - m)q_{t,1}' + mq_{t,2}'$$

$$q_{t+1,2} = (1 - m)q_{t,2}' + mq_{t,1}'$$

Once again let us suppose that this process is repeated until a stable equilibrium is reached. Due to the assumed symmetry of the model, we know that any equilibrium at which both behaviors are present must satisfy

(9) $\hat{q}_1 = 1 - \hat{q}_2$

Where $\hat{q}_1$ is the fraction of individuals acquiring behavior 1 in environment 1, and $\hat{q}_2$ is the fraction of individuals acquiring behavior 1 in environment 2. Using this fact one can show that

(10) $$\hat{q}_1 = \frac{(1 - 2m)p_{11} + m}{(1 - 2m)(p_{11} + p_{12}) + 2m}$$

Notice that when $m = 0$, (10) reduces to the equilibrium derived in the model without any environmental variation. Also notice that if individuals are equally likely to imitate models drawn from both environments (i.e. $m = 1/2$), then $\hat{q}_1 = 1/2$. For intermediate values of m, $\hat{q}_1$ falls between these two extreme values.

    These properties make sense. In a uniform environment the behavior that results in higher fitness will increase in frequency according to the simplified model of the previous section; individuals should depend entirely on social learning and not take a chance on trial and error learning. When $m = 0$, there is no contact between individuals who experience the different environments, and the correct behavior in each environment becomes overwhelmingly common. Individual learning cannot do better than a perfected tradition, and it will frequently lead to errors. Within-cohort environmental variation, represented now by the movement of individuals among groups exposed to different environments, causes individuals to be exposed to some immigrant models who are likely to have acquired the behavior favored by individual learning in the other environment. Therefore, the movement of models among groups in a spatially variable environment causes social learning to be a less reliable method of acquiring one's behavior than it is in a homogeneous environment. When $m = 1/2$ the frequency of the superior behavior is increased in each environment by the effects of individuals' experience, but the mixing of models from the two environments exactly erases the gains, and the individuals in the next cohort must start from scratch. In this case social learning is useless.

The most interesting cases are the ones at intermediate values of m where both social and individual learning are likely to be important. We will now compute the ESS amount of social learning in a variable environment for $0 < m < 1/2$. The expected fitness of individuals using a learning rule characterized by the learning parameter d' is given by

(11) $E\{w(d')\} = W + D[\hat{q}_1(d)(1 - p_{11}(d') - p_{12}(d')) + p_{11}(d')]$

where $\hat{q}_1(d)$ is the equilibrium frequency of trait one in habitat 1 assuming that virtually all of the population is characterized by learning parameter d. To determine the ESS value of d, the confidence-interval-like parameter that determines the relative importance of social and individual learning, we once again determine which value of d can resist invasion by modifying alleles. A population in which d predominates can resist invaders which increase d whenever:

(12)

$$(1\text{-}2m) \left[\frac{\partial p_{11}}{\partial d} p_{12}(d) - \frac{\partial p_{12}}{\partial d} p_{11}(d) \right] + m \left[\frac{\partial p_{11}}{\partial d} - \frac{\partial p_{12}}{\partial d} \right] < 0$$

Consider how varying d affects the sign of the left hand side of (12). We know from the models of a constant environment that the first on the left hand side of (12) is always positive (see eqn. 5). It is clear from the definition of p_{11} and p_{12} (see figure 1) that the second term equals zero when $d = 0$, and is negative for all larger values of d. This means that when $d = 0$, the left hand side of (12) will be positive and alleles which increase d can invade. Next notice that as d becomes large both p_{11} and p_{12} approach zero, and therefore for large enough values of d, the left hand side of (12) is negative, and alleles which decrease d can invade. Taken together these facts mean that expected fitness is maximized for some amount of social learning intermediate between zero and one as long as $1/2 > m > 0$. While we have not been able to solve (12) analytically, it is easy to solve numerically. The results, shown in figures 2 and 3, suggest that under a wide combination of migration rates and quality of individual experience, it is optimal to employ a mixture of social and individual learning. There is a broad region with combinations of modest migration rates and moderate to low information quality where social learning should be rather more important than individual learning in determining individual behavior. In figure 2, the ESS value of d, d^*, is plotted as a function of S, the measure of the quality of the information available to individuals, and the probability that naive individuals are exposed to models who learned from the wrong environment (m). There are two things to notice about these results: first, as

individual experience becomes less reliable (i.e. S becomes large)
the optimal amount of social learning is increased. Second, as
the environment becomes less predictable (i.e. m increases) the
optimal amount of social learning decreases. In figure 3, we plot
the probability that individuals rely on social learning
($L^* = 1 - p_{11}(d^*) - p_{12}(d^*)$) given that d equals its optimal
value.

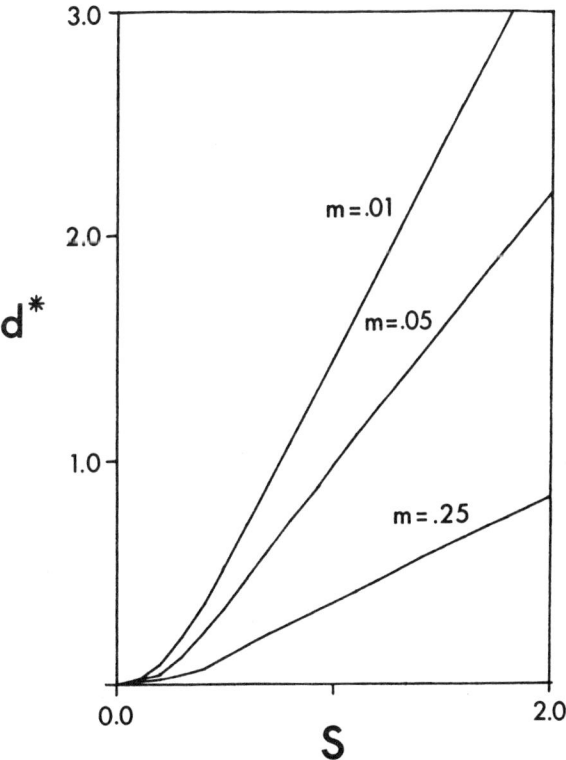

Figure 2: Plots the evolutionary equilibrium value of d, d^*,
as a function the quality of information available for individual
learning, S, and for three levels of environmental heterogeneity,
measured by m.

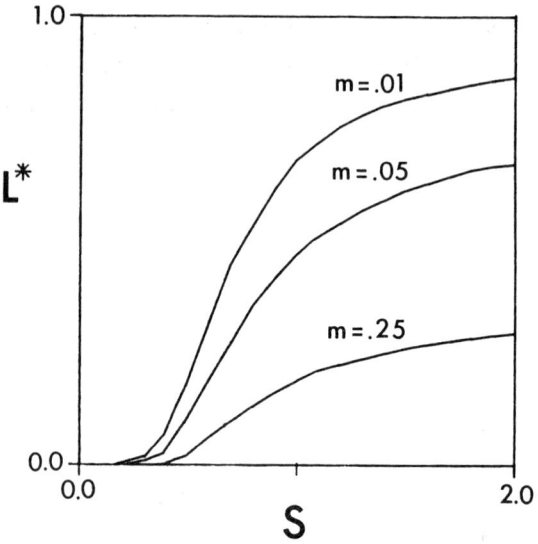

Figure 3. Plots the fraction of the population acquiring behavior by social learning when d is at its equilibrium value, $L^* = (1 - p_1(d^*) - p_2(d^*))$, as a function of S and m.

This model suggests that the adaptiveness of social learning relative to individual learning depends on two factors, the accuracy of individual learning, and the chance than an individual's social models experienced the same environment that the individual experiences. A substantial dependence upon social learning seems to be most adaptive when individual learning is inaccurate, and there is not too much migration among habitats. The occasional use of individually-acquired compelling evidence, coupled with faithful copying in the absence of such evidence, is sufficient to keep the locally adaptive behavior common.

Increasing the importance of individual learning would entail more errors and would reduce the frequency of the adaptive behavior. In contrast, when there is extensive migration among habitats, relatively rare instances of individual learning would not be sufficient to maintain a high frequency of the locally adaptive behavior. Under such conditions, individuals must rely on individual learning if they are to have any chance of acquiring locally adaptive behavior.

Similar results derived using different models suggest that the present conclusions are robust. We have analyzed the same dichotomous model in a temporally fluctuating environment (Boyd and Richerson, 1988). Assuming a Markov model of environmental change, we showed that the ESS reliance on social learning has the same qualitative properties as the model analyzed here. Elsewhere (Boyd and Richerson, 1985:ch. 4) we have analyzed a model which embodies the same qualitative assumptions about the nature of social learning and individual learning but in which behaviors are formalized as quantitative characters. These models have the same qualitative conditions for the evolution of social learning that result from the present model. Finally, we have also extended the analysis of these models to allow for the genetic transmission of behavioral predispositions in addition to the genes which affect learning (Boyd and Richerson, 1983, 1985:ch 4).

5. SOCIAL LEARNING WITH MORE THAN ONE MODEL. One can think of social learning as using the behavior of others as a source of information about the environment. Adaptive processes such as individual learning will often cause the more common behavior to also be the most adaptive behavior, and, therefore, copying the behavior of a randomly chosen individual can be adaptive under the right circumstances. In many species, however, naive individuals may be able to observe the behavior of a number of experienced conspecifics. That is, each naive individual often has a set of

models.   When this is the case, one can think of such sets of
models as samples of the behavior present in the population.   Then
if there is behavioral variation in a population, different
individuals will be exposed to different samples of that behavior.
Since different samples of behavior lead to different inferences
about the commonness of one or the other behaviors in the
population, it seems plausible that naive individuals exposed to
different samples of behavior might differ in the extent to which
they rely on social learning versus individual learning.

To address this question, we have modified the model
presented above so that individual are exposed to the behavior of
n models.   Now it may be the case that an individual's models
differ in their behavior and the naive individual is confronted
with the problem of deciding which variant to adopt. There is also
an opportunity afforded by a large set of models. Since individual
learning will tend to increase the frequency of behaviors that are
adaptive in a the local habitat, there may well be information in
the model "sample" as to what behaviors are adaptive, especially
as the size of the sample of the previous generation increases.
Selection might structure social learning so as to use this
information.   We want to determine the evolutionarily stable
solutions to this problem.

Begin by considering an individual exposed to i models using
behavior 1 and $n - i$ models using behavior 2.   Once again assume
that the individual observes the variable x that indicates the
state of the environment and then adopts each behavior with the
probabilities given in table 2.   As before the value of d_i
determines the minimum quality of information necessary before the
individual will rely on individual learning.   It is indexed by i
to indicate that individuals may have different thresholds
depending on the number of models who use one behavior or the
other.   We further assume that d_i = d_{n-i}. This assumption
formalizes the idea that it is the *number* of models who use a

given behavior that governs the usefulness of information acquired by social learning, not which trait they use. The value of A_i determines the conditional probability that the individual will acquire behavior 1 given that it is going to rely on social learning. To represent the idea that there is no innate predisposition to adopt either trait in the absence of information about the environment, we assume that $A_i = 1 - A_{n-i}$.

| | Probability of acquiring | |
Event	Behavior 1	Behavior 2
$d_i < x$	1	0
$-d_i < x < d_i$	A_i	$(1 - A_i)$
$x < -d_i$	0	1

Table 2.

As before suppose that there are two habitats linked by migration, one in which behavior 1 is favored and one in which behavior 2 is favored. Let the frequency of the behavior 1 among models in environment j be $q_{t,j}$. Further suppose that models are sampled at random from the population. With these assumptions the frequency of behavior 1 in environment j after individual and social learning, $q_{t,j}'$, is

$$(14) \qquad q'_{t,j} = \sum_{i=1}^{n} \binom{n}{i} q_{t,j}^{i} (1 - q_{t,j})^{n-i}$$
$$\{A_i(1 - p_{j1}(d_i) - p_{j2}(d_i)) + p_{j1}(d_i)\}$$

The frequencies in each habitat after migration are given by equation (8).

The next step is to determine the equilibrium frequency of behavior 1 in each habitat. Because equation (14) is quite complex have not been able to derive an analytical expression for these equilibrium frequencies. However, it follows from the symmetry of the model that there is a stable symmetric equilibrium such that the favored behavior is common in each habitat, i.e., $\hat{q}_1 = 1 - \hat{q}_2 > 1/2$. We will refer to this as the symmetric equilibrium. Depending on the values of A_i and d_i there may also be other stable internal equilibria at which one behavior is common in both habitats.

To determine evolutionarily stable pattern of social learning, assume that most of the population has a learning rule characterized by the sets of parameters $d = \{d_0,..,d_n\}$ and $A = \{A_0,..,A_n\}$, and that the population has reached the resulting symmetric equilibrium. Then an individual with a different learning rule characterized by the sets of parameters $d' = \{d_0',..,d_n'\}$ and $A' = \{A_0',..,A_n'\}$, has expected fitness given by

$$E\{w(d',A')\} = W + D \sum_{i=0}^{n} \binom{n}{i} \hat{q}_1^i (1 - \hat{q}_1)^{n-i}$$

(15)
$$\{A_i'(1 - p_{11}(d_i') - p_{12}(d_i')) + p_{11}(d_i')\}$$

where $\hat{q}_1$ is the frequency of the favored behavior in each habitat at the symmetric equilibrium resulting from A and d. Then using the fact that $A_i = 1 - A_{n-i}$ and $d_i = d_{n-i}$, it can be shown that alleles which lead to a small increase in A_i can invade if:

(16)
$$\hat{q}_1^i (1 - \hat{q}_1)^{n-i} - \hat{q}_1^{n-i} (1 - \hat{q}_1)^i > 0$$

which is always satisfied for $i > n/2$. Thus the ESS values of A_i, A_i^*, are given by:

(17)
$$A_i^* = \begin{cases} 1 & i > n/2 \\ 1/2 & i = n/2 \\ 0 & i < n/2 \end{cases}$$

This says that given that an individual is going to rely on social learning, it should always adopt the more common behavior exhibited by its models. At the symmetric equilibrium the favored behavior is more common in each habitat. Thus, if individual experience is not determinative, the best thing to do is copy the behavior that is most common among models as it is more likely to be the locally favored behavior.

To determine the ESS value of d_i, d_i^*, assume that the set of A_i are at their ESS values given by (17). Then alleles which lead to a small increase in d_i can invade if

(18) $$\frac{\partial p_{12}}{\partial d_i} \, \hat{q}_1^i (1 - \hat{q}_1)^{n-i} - \frac{\partial p_{11}}{\partial d_i} \, \hat{q}_1^{n-i} (1 - \hat{q}_1)^i > 0$$

Substituting the definitions of p_{11} and p_{12} and simplifying yields the following expression for the ESS value of d_i

(19) $$d_i^* = (S/M)(n/2 - i)\{\ln\hat{q}_1 - \ln(1 - \hat{q}_1)\}$$

This expression says that when an equal number of models use each behavior ($i = n/2$), individuals should ignore their models and rely completely on individual learning. As the number of models exhibiting one behavior increases, d_i^* also increases linearly, and therefore the relative importance of individual learning declines. This effect becomes stronger as the frequency of the favored behavior in each habitat increases and as the size of the set of models increases. When nearly everyone in a given habitat uses the optimal trait and your set of models gives clear indication which behavior is more common in the local habitat, then you should only adopt the alternative behavior if the evidence from your own experience is very strong. On the other hand, if both behaviors almost equally common in both habitats, the fact that one behavior is common among your models gives little information about the which behavior is favored locally (especially if the number of models is small), and individuals should mainly rely on their own experience.

6. DISCUSSION. The models presented in this paper lead to three qualitative conclusions about the evolution of social learning. First, the adaptiveness of social learning depends on a tradeoff. Increasing the importance of social learning increases fitness because it allows a reduction in the error rate of individual learning. However, increasing the importance of social learning also decreases the ability of the population to track a variable environment. A heavy dependence on social learning relative to individual learning seems to be most adaptive when (1) individual learning is error prone, and (2) environments are predictable. Second, the models suggest that when individuals do depend on social learning in a variable environment, they should not imitate randomly chosen individuals. Rather, they should tend to imitate the more common behavior among their models. This result follows from the fact that the behaviors favored by selection in a particular environment will tend to be more common in that environment. Finally, the models presented here suggest that selection will favor a pattern of social learning in which individuals exposed to more variable sets of models rely more heavily on individual learning. Given that models are numerous and sampled at random from the population, a predominance of one behavior among the models indicates that that behavior is more common in the population from which the models were drawn, and therefore, likely to be adaptive. An even mix of behavior among models indicates little about which behavior is common, especially if the number of models is small. Therefore, it may make sense to depend heavily on individual learning.

The models presented in this paper can be thought of as a generalization of statistical decision theory. Within the context of that body of theory, decision makers seek to choose the best decision from among set of possibilities given specified information about the relationship between alternative decisions and outcomes. While this information may be imperfect, its statistical properties are specified, and they are independent of the decisions made by others. Given these assumptions it is possible to specify the best decision procedures by considering each decision maker in isolation. Social learning involves decision makers who use the behavior of others as part of the information on which they base their decisions. The behavior of others depends on the decisions those individuals made, and therefore their decision rules. To specify the best rules for social learning one must determine how a given decision rule affects the distribution of observed behavior in a population of decision makers. The models presented here provide one simple example of how this might be done in the context of the evolution of social learning.

The models presented here are very general, and should apply to many situations in which animals could get information about the environment by observing conspecifics. The apparent rarity, or at least lack of sophistication, of social learning in species besides humans (Galef, 1988) is a considerable puzzle given our results. Understanding the adaptive properties of social learning presents an array of fascinating theoretical and empirical problems.

BIBLIOGRAPHY

1. Boyd, R. & P. J. Richerson "The cultural transmission of acquired variation: effect on genetic fitness," J. Theor. Biol. 100 (1983)567-596.

2. Boyd, R. & P. J. Richerson, Culture and the Evolutionary Process, University of Chicago Press, Chicago, 1985.

3. Boyd, R. & P. J. Richerson, "The evolution of cultural transmission: The effects of spatial and temporal variation," In: Social Learning: A Biopsychological Approach, T. Zentall and B. G. Galef, eds. pp. 29-48, Lawrence Erlbaum Associates, Hillsdale, NJ, 1988.

4. Galef, B. G. "Social transmission of acquired behavior: A discussion of tradition and social learning in vertebrates," In: Advances in the Study of Behavior, Vol. 6, J. S. Rosenblatt, R. A. Hinde, E. Shaw, and C. Beer, eds., pp. 77-100. Academic Press, New York, 1976.

5. Galef, B. G. "Imitation in animals: History, definiton, and interpretation of data from the psychological laboratory," In: Social Learning: A Biopsychological Approach. T. Zentall and B. G. Galef, eds. pp. 1-28, Lawrence Erlbaum Associates, Hillsdale, NJ, 1988.

6. Hauser, M. "Invention and social tranmission: New data from wild vervet monkeys." In: Machiavellian Intelligence: Social Expertise and the Evolution of Intellect in Monkeys, Apes, and Men. R. W. Byrne and A. Whitten eds. pp. 327-344, Clarendon Press, Oxford, 1988.

7. McNamara, J. & A. Houston, 1980. "The application of statistical decision theory to animal behavior," J. Theor. Biol. 85 (1980) 673-690.

8. Marler, P. & M. Tamura, "Culturally transmitted patterns of vocal behavior in sparrows," Science, 146 (1964) 1483-1486.

9. McGrew, W. C. & C. E. G. Tutin, "Evidence for a social custom in wild chimpanzees?" Man, 234 (1978) 234-251.

10. Staddon, J. Adaptive Behavior and Learning, Cambridge University Press, Cambridge, 1983.

11. Stephens, D. and Krebs, J. Foraging Theory. Princeton University Press, Princeton, NJ, 1987.

DEPARTMENT OF ANTHROPOLOGY
UNIVERSITY OF CALIFORNIA
LOS ANGELES, CA 90024

INSTITUTE OF ECOLOGY
UNIVERSITY OF CALIFORNIA
DAVIS, CA 95616

Lectures on Mathematics in the Life Sciences
Volume 20, 1989

DETERMINISTIC MULTILOCUS POPULATION GENETICS: AN OVERVIEW

ALAN HASTINGS[1]

ABSTRACT. The models of deterministic multilocus population genetics are reviewed. Emphasis is placed on delineating regions of parameter space, interms of the strength of selection, where behavior of the multilocus models is simple. The kinds of complicated behavior possible when selection is neither weak nor strong relative to recombination are reviewed.

1. INTRODUCTION

Population genetics is one of the fields of biology where mathematics has been most successfully applied (see, for example the review in Ewens, 1979). I will not attempt an exhaustive review here, nor will I define many of the basic terms of population genetics, which can be found in many basic texts (e.g. Hartl 1980). The book by Ewens and the more elementary texts it references provide further background. My goal here is to concentrate on one particular aspect of population genetics theory, the study of viability selection at multiple loci. I also discuss only the deterministic aspects of these models, ignoring the vast and important literature on stochastic models.

1980 Mathematics Subject Classification (1985 Revision). 92A10, 92A12.
Supported by grant GM32130 from the Public Health Service.

The basic questions addressed by population genetics lie within several areas. These questions all overlap to some degree, but I will delineate several areas. One is determining the factors that maintain maintains variability in natural populations. Any individual has many (thousands) of genes, which in diploids, are represented by two copies. The sites at which genes occur are referred to as *loci*, and the different possible genes at a site are called *alleles*. Observations of natural populations at the level of the single locus (see for example summaries in Lewontin, 1974 and Nei, 1987) show that often a number of different alleles are present in appreciable frequencies.

Another aspect of this question can be phrased at the level of the *phenotype*, the trait exhibited by the individual, which is determined jointly by the genotype (genetic makeup) and the environment (see Fisher, 1918 and Falconer, 1981). For many characters which vary (almost) continuously, such as height, many genes contribute to the phenotype. Variability referred to as additive genetic variability, is maintained in natural populations. Why are all types which deviate from an 'optimal' genetic type not eliminated ? A number of answers to this question have been proposed (Turelli, 1984). Most theoretical approaches to this question have made a number of assumptions, in particular assuming that linkage can be ignored, so I will not address these questions further (Turelli, 1984).

A second question, partly, dependent on the first, concerns the response of populations to the effects of selection. What is the response of populations to selection, either natural, or artificial ? How fast will phenotypes change, and what will be the underlying changes in the genetic composition ?

A third question, also related to the first two, takes the viewpoint of the individual locus. Changes in alleles occur over long times, and can be observed at the level of individual amino acids (see summary in Kimura, 1983). What are the mechanisms responsible for causing these changes, and what is the relationship of changes at this level to changes at the level of the phenotype?

Although population genetics is much more complex than what I have just outlined, the major questions usually fall into the areas described above. Moreover, there is a well developed mathematical theory that has helped to guide current thinking regarding these questions. This theory had its beginnings in the work of three giants, Fisher, Haldane and Wright (see Provine, 1971 and Ewens, 1979), whose work shaped the field of theoretical population genetics.

The early theory of population genetics for the most part concentrated on questions concerning the behavior of single loci. This is particularly true of the work of Fisher and Haldane. One of the most celebrated results in population genetics, Fisher's Fundamental theorem, which states that mean fitness always increases only holds for single loci. Attempts to extend this theory to multiple loci are discussed below.

Wright, on the other hand, in his verbal theories emphasized the role of interactions among genes at different loci. His 'shifting balance theory of evolution' was in fact a theory that required the inclusion of multilocus dynamics. The inclusion of multiple loci is difficult because not only do fitnesses depend on many loci, but also because it becomes important to consider statistical correlations for the joint occurrence of alleles at different loci, which has been referred to as disequilibrium. Wright did begin the consideration of multiple locus models in genetics. He first considered approaches which did not include disequilibrium and then considered some of the difficulties introduced by explicitly including disequilibrium. This theory was carried on in some early papers by Kimura and Lewontin and Bodmer and Felsenstein. This work was then extended in a variety of directions by these and a number of other authors, notably Ewens, Feldman, Karlin, McGregor and Nagylaki.

I will now turn to a more systematic development. Previous reviews of multilocus population genetics theory are in Karlin (1975), Karlin (1978), Ewens (1979). The theme around which this review is organized is the following. The single locus models of population genetics have fairly simple behavior because the fitness acts as a Liapunov function. On a deeper level, these systems are simple because they are gradient systems (at least the continuous time versions are), as discussed by Akin (1979), building on work of Shashahani (1979). Thus no cycles are possible, and all asymptotic states are equilibria. Also, as noted explicitly by Karlin and MacGregor (1972), if recombination is small, or in other words selection is strong, then one locus results provide a good guide to the behavior of multilocus systems. Further, if linkage equilibrium is assumed (statistical correlations among alleles at different loci are ignored), then once again the (continuous time versions of the) models of population genetics become gradient

systems, as was implicit in the work of Wright (1935) and again explored in
mathematical detail by Akin (1979). One can show that linkage equilibrium holds
approximately if selection is weak, relative to recombination. Thus both for weak
and for strong selection, simple behavior is expected for population genetic models.
Below, I will explore to what extent more complex behavior is possible in these
models for the cases where selection and recombination have roughly the same
magnitude. Three natural questions will emerge: (1) What is the range of behavior
possible ? (2) For what combinations of selection strength and recombination do
population genetic models exhibit more complex behavior? (3) Independent of the
strength of recombination or selection, are there patterns of selection for which the
simple asymptotic results for one locus or weak selection hold?

2. SINGLE LOCUS THEORY

I will first summarize very briefly the relatively complete theory of selection
at a single locus. The first reason for this is to place the questions of multilocus
theory in a proper context. The second is that there are at least two circumstances in
which single locus theory provides a guide to multilocus theory. One is the case
where recombination is very small. If there is no recombination, then multilocus
models reduce to single locus models. Thus, if the recombination rate is small, one
can determine the location and stability of equilibria for a multilocus model from a
single locus model. The mathematical theory for this is developed in Karlin and
MacGregor (1972) and the use of this approach in multilocus population genetics is
reviewed in Karlin (1978). A second case where single locus theory provides a
guide is if disequilibrium, or deviations from random combination, is ignored (e.g.
Nagylaki, 1976, Akin, 1979). Again, in this case, one can infer dynamic and static
properties from single locus results.

In a single locus model, let there be n alleles, numbered 1 through n. Let
the frequency of allele i be given by p_i. Let the fitness of an individual with alleles i
and j be $w_{ij}=w_{ji}$. Then the dynamics of this system in discrete time are given by:

$$p_i' = p_i w_i / \bar{w},$$
(1)

where the prime denotes frequencies in the next generation. Here

$$w_i = \sum_j p_j w_{ij} \tag{2}$$

is the mean fitness of allele i, and

$$\bar{w} = \sum_j p_j w_j \tag{3}$$

is the mean fitness of the population. The dynamics of the system (1)-(3) are quite well understood because the mean fitness acts like a Liapunov function, as in the following result, which is demonstrated elegantly in Kingman (1961).

Theorem. The mean fitness, $\bar{w}$, satisfies the following for the model (1)-(3):

$$\bar{w}(t+1) \geq \bar{w}(t), \tag{4}$$

with equality if and only if

$$p_i(t+1) = p_i(t) \quad \forall \ i. \tag{5}$$

A second result in Kingman (1961) provides an elegant characterization of when a stable equilibrium exists.

As one important consequence of these results the following is a necessary and sufficient condition for a stable polymorphism with two alleles:

$$w_{12} \geq w_{11}, w_{22}, \tag{6}$$

i.e. the heterozygote is more fit than either homozygote. The condition (6) is referred to as either overdominance or heterosis. Similar conditions can be given for more alleles (Kingman, 1961), but do not lend themselves to as simple an interpretation.

3. QUESTIONS OF MULTILOCUS THEORY

Many of the questions of multilocus theory can be subsumed under the general question of how important is disequilibrium. As noted above, if disequilibrium can be ignored, then the equations for multiple loci reduce to a from which can be treated by many of the methods of single loci. One reasonable approach would be to determine the importance of disequilibrium through experimental observation, to determine how large and how common are deviations from random combination in nature. Unfortunately, this is a difficult problem statistically (see Brown, 1975), as it is difficult to reject the null hypothesis of random assortment. This may explain why deviations from random combination have not been detected frequently except in inbreeding species (see review in Hedrick et. al., 1978), a topic which will be discussed below.

Thus one major question is how large is disequilibrium at equilibrium? This is important for many reasons. One reason for the importance of this question is the role played by disequilibrium in modifier theory (e.g. Feldman and Liberman, 1986), which is the study of loci at which the alleles affect mutation, recombination, traits which affect the dynamics of the genetic system. I do not pursue this interesting subject further.

A second series of questions relates to the nice properties of one locus models, which result from properties of the mean fitness. In one locus models, there is at most one polymorphism with a given set of alleles. The only attracting sets are equilibria -- cycles or more complicated dynamic behavior is impossible. Mean fitness always increases. How many of these properties carry over to multiple locus questions and what form do they take for multiple loci?

A third series of questions is related to the first concerning the role of disequilibrium. Observations of gene frequencies in populations, of necessity, concentrate only on some of the loci in the organisms. How do these marginal systems behave? In particular, what are the differences between the behavior of 'real' one locus systems and marginal one locus systems?

The definition of marginal genetic systems (Ewens and Thomson, 1977) will be useful both in answering these questions and in the analysis of multilocus models. Suppose the fitnesses depend on n loci when only m loci are in fact observed. Define a marginal m-locus subsystem of a n-locus system as the system obtained by averaging all the fitnesses over the missing loci, weighted by the appropriate frequencies of the gametes. Thus, following Ewens and Thomson, denote the frequency of the n-locus gamete i by $x(i)$ and the frequency of the m-locus gamete p by $z(p)$. By properly renumbering the loci the m loci being considered can be thought of as being the ones numbered one through m in the full n-locus system. (Note that this means that the numerical order of the loci need not correspond to the physical order on the chromosome). Thus

$$z(p) = \sum_{i \in S_p} x(i) . \tag{7}$$

where the set S_p is defined as:

$$S_p = (i \mid i_k = p_k, k=1,m), \tag{8}$$

the set of n locus gametes i that have the same alleles at the first m loci as the m locus gamete p.

The induced marginal fitness of the genotype formed by the m-locus gametes p and q will be denoted by $\bar{w}_{pq}$ and is obtained by averaging over all genotypic combinations making up these two m-locus gametes, weighted appropriately by fitnesses and frequencies. This yields:

$$\overline{w}_{pq} = \sum_{i \in S_p, j \in S_q} x(i)x(j)w_{ij}/(z(p)z(q)). \tag{9}$$

This definition is useful below because of the following fact found by Ewens and Thomson. The dynamic equations for the m- locus subsystem are the same as those for a genuine m-locus system, with the marginal fitnesses taking the place of the actual fitnesses. Note, however, that the marginal fitnesses are not constants, but depend on allele frequencies and disequilibria at other loci. Thus, the marginal systems are particularly important for deducing equilibrium behavior. A slight change in notation will prove convenient below. Since the gametes themselves denote the number of loci being considered, the 'x' designation for marginal systems will be used at times instead of the 'z' designation. No confusion results, since the number of loci being specified is given explicitly.

4. TWO LOCUS MODELS

I will now begin a review of attempts to answer these questions. As an indication of the difficulty of the problems, results for two loci are much better developed than results for more loci, so I will begin with a discussion of results for two loci. Even for two loci, a complete theory of the behavior of deterministic models has not yet been possible, so I will first deal with special models, and then describe more general results. As a preliminary step, I will describe the standard two locus two allele, deterministic discrete time model.

The model is based on the standard models of two locus theory (see, e.g. Ewens, 1979). I will first describe the two locus model. Let there be two loci with two alleles each: A and a at the A locus and B and b at the B locus. Let the frequency of the gametes AB, Ab, aB, ab be given by x_1, x_2, x_3, x_4, respectively. Define the linkage disequilibrium D to be $x_1x_4 - x_2x_3$. Let r be the recombination rate, the probability that an individual with alleles A and B on one chromosome, and a and b on the other produces a gamete of type Ab or aB. Let δ_i take the values -1,1,1,-1 for i =1 to 4 respectively. Let w_{ij} be the fitness of the individual with the gametes whose frequencies are x_i and x_j. Note that $w_{ij}=w_{ji}$. Also, if the fitness is independent of the gametes on which the alleles occur, i.e. there are no cis-trans effects (see Turelli, 1982), one can assume that $w_{14} = w_{23} = 1$. Under this assumption, the fitnessses can be displayed conveniently as the following three by three matrix:

$$
\begin{array}{ccc}
\text{BB} & \text{Bb} & \text{bb} \\
\end{array}
$$

$$
\begin{array}{c}
\text{AA} \\
\text{Aa} \\
\text{aa}
\end{array}
\left[
\begin{array}{ccc}
w_{11} & w_{12} & w_{22} \\
w_{13} & w_{23}=w_{14} & w_{24} \\
w_{33} & w_{34} & w_{44}
\end{array}
\right] \tag{10}
$$

Denote the marginal mean fitness of the gamete i by:

$$
w_i = \sum_{j=1}^{4} x_j w_{ij}, \tag{11}
$$

and the mean fitness of the population by:

$$
\bar{w} = \sum_{j=1}^{4} x_j w_j, \tag{12}
$$

Then the dynamics of this system are given by:

$$x_i' = (x_i w_i + \delta_i rD)/\overline{w} \tag{13}$$

for a discrete time system with nonoverlapping generations.

Additive Model

In one special model, fitnesses are assumed to be determined additively from fitnesses at individual loci. This is easiest to see if the fitnesses just introduced are written as the 3 x 3 matrix described above. For the *additive* model, fitnesses written in the form (10) are given by:

$$
\begin{array}{ccc}
a_0 + b_0 & a_0 + b_1 & a_0 + b_2 \\
a_1 + b_0 & a_1 + b_1 & a_1 + b_2 \\
a_2 + b_0 & a_2 + b_1 & a_2 + b_2
\end{array}
\tag{14}
$$

I will now describe the behavior of two locus additive models. First, assume that the fitnesses satisfy the conditions:

$$a_1 > a_0, a_2 \tag{15}$$

$$b_1 > b_0, b_1 \tag{16}$$

so that each locus in isolation would have a polymorphic equilibrium. In this case the model (13) has a unique polymorphic equilibrium, with, for example, the frequency of the gamete containing alleles A and B given by:

$$[(a_1 - a_2)/(2a_1 - a_0 - a_2)][(b_1 - b_2)/(2b_1 - b_0 - b_2)], \tag{17}$$

which is the product of the one locus equilibrium frequencies. All of the other equilibrium gamete frequencies are similar. Because of the form of this equilibrium, it will be called the product equilibrium. Ewens (1969) proved that with fitness given by (14), the mean fitness $\bar{w}$ for is nondecreasing. Karlin and Feldman (1970a) made use of this result to show that the equilibrium described by (17) is in fact globally stable -- it is approached by all starting conditions with all alleles initially present. Unfortunately, the simple behavior of the additive model is the exception.

Multiplicative Model

Another simple possibility is the multiplicative model, which implies another kind of independence in the determination of fitnesses. Instead of assuming that fitnesses are determined by adding contributions at different loci, one can assume multiply contributions at different loci, leading to the following fitness matrix displayed in the form given in (10):

$$
\begin{array}{lll}
a_0 b_0 & a_0 b_1 & a_0 b_2 \\
\\
a_1 b_0 & a_1 b_1 & a_1 b_2 \\
\\
a_2 b_0 & a_2 b_1 & a_2 b_2
\end{array}
\tag{18}
$$

Both the additive and multiplicative models imply a lack of *epistasis*, fitnesses do not depend on interactions between alleles at different loci. This is easily seen if a compact notation developed by Karlin and his coworkers(see for example the review in Karlin, 1978) is used. This notation is also essential for the development of most analytic results for more than two loci. Denote by $A \otimes B$ the Kronecker (or outer) product of the matrices A and B (see Karlin, 1978). Denote by $u \circ v$ the Schurr product of the vectors u and v, so $(u \circ v)_i = u_i v_i$. Let

$$M_1 = \begin{bmatrix} a_0 & a_1 \\ a_1 & a_2 \end{bmatrix} \tag{19}$$

$$M_2 = \begin{bmatrix} b_0 & b_1 \\ b_1 & b_2 \end{bmatrix} \tag{20}$$

$$E = \begin{bmatrix} 1 & 1 \\ 1 & 1 \end{bmatrix} \tag{21}$$

Then, where I is the identity matrix, one can write equation (13) describing the two locus model, for the fitnesses given in (18) as:

$$W(x)x' = (1-r)(I \otimes I)x \circ (M_1 \otimes M_2)x + r(M_1 \otimes I)x \circ (I \otimes M_2)x, \tag{22}$$

where

$$W(x) = (x, M_1 \otimes M_2 x). \tag{23}$$

Also, equation (13) can be written as, for the fitness given in (14):

$$W(x)x' = (1-r)(I \otimes I)x \circ (M_1 \otimes E)x + r(M_1 \otimes I)x \circ (I \otimes E)x$$
$$+ (1-r)(I \otimes I)x \circ (E \otimes M_2)x + r(E \otimes I)x \circ (I \otimes M_2)x, \tag{24}$$

where

$$W(x) = (x, M_1 \otimes Ex) + (x, E \otimes M_2 x). \tag{25}$$

Thus, in this notation, $x(i)$ is a polymorphic equilibrium for the one locus fitness matrix M_i if

$$M_i x(i) = u (M_i x(i), x(i)), \tag{26}$$

where u is the vector of all ones. The product equilibrium defined above becomes

$$x^* = x(1) \otimes x(2). \tag{27}$$

Substituting using standard properties of Kronecker products, it is straightforward to see that both equations (22) and (24) always admit the product equilibrium (27). The local stability analysis of the product equilibrium for both models is facilitated by the properties of the eigenvalues and eigenvectors of Kronecker products. This analysis (e.g. Karlin, 1978) yields the following result (obtained earlier by direct computation). Assume that the fitnesses for the multiplicative model are such that each locus in isolation would have a polymorphism. Then for large enough values of the recombination rate r, the product equilibrium is stable, and it is unstable for smaller values of r.

Symmetric Model

Another special fitness matrix is the symmetric fitness form, which can have large epistatic effects. A special case of this was introduced by Wright (reviewed in Wright, 1969) because of the connection with models of quantitative genetics. In its most general form, the fitnesses for the symmetric model, again displayed in the form (10) are:

$$\alpha \quad \gamma \quad \beta$$

$$\delta \quad 1 \quad \delta \tag{28}$$

$$\beta \quad \gamma \quad \alpha$$

A relatively complete study of the equilibrium structure of the model (28) was completed by Karlin and Feldman (1970b). A summary of their work which built upon earlier work by Lewontin and Kojima (1960) and Ewens(1969), among others, follows.

In the Lewontin-Kojima version of the symmetric viability model one assumes that the fitnesses of all the homozygotes are the same. So, let

$$w_{11} = w_{22} = w_{33} = w_{44} = 1 - a. \tag{29}$$

Additionally it is assumed that

$$w_{12} = w_{34} = 1 - b \tag{30}$$

and

$$w_{13} = w_{24} = 1 - g. \tag{31}$$

Here a, b, g are assumed positive and represent the strength of selection against homozygotes and two kinds of single heterozygotes respectively.

This model has been extensively studied (Karlin, 1975 and Ewens, 1979 provide reviews). There is always a product equilibrium of the form:

$$x_1 = x_2 = x_3 = x_4 = 1/4, \tag{32}$$

which is always unstable if

$$|b - g| > a. \tag{33}$$

Additionally, if

$$r < (b + g - a)/4 \tag{34}$$

there are two 'symmetric' equilibria equilibria of the form:

$$x_1 = x_4 = (1 + f)/4$$
$$\tag{35}$$
$$x_2 = x_3 = (1 - f)/4$$

and

$$x_1 = x_4 = (1 - f)/4$$
$$\tag{36}$$
$$x_2 = x_3 = (1 + f)/4$$

where

$$f = [1 - 4r/(b+g-a)]^{1/2}. \tag{37}$$

Ewens (1968) showed that these equilibria are stable if and only if

$$4r^2(b+g-a)+2r(2a^2-b^2-g^2-a(b+g))+a(b+g-a)^2 > 0. \tag{38}$$

Karlin and Feldman (1970b) discovered the existence of up to four further 'unsymmetric' equilibria and provided formulas for them. They gave a detailed analysis, including a stability analysis, for the special case b=g. Under the conditions they analyzed, no more than two simultaneously stable equilibria are possible.

In Hastings (1985), I used bifurcation theory to provide a partial analysis of the unsymmetric equilibria in the case not studied in detail by Karlin and Feldman, namely b≠g (subject to the restriction that the homozygotes all have the same fitness) using techniques developed in Hastings (1982). The results given by bifurcation theory are not as complete as those obtained by an exact determination of the equilibria, but the algebra is much simpler and the stability of the equilibria follows immediately. The analysis is based on an analysis of the corner equilibrium (see Bodmer and Felsenstein, 1967 for preliminary calculations on the stability of this equilibrium)

$$p_A = p_B = 0, \tag{39}$$

where $p_A = x_1 + x_2$ is the frequency of allele A and p_B is defined similarly. Note that because of the symmetry assumed here, the same analysis is valid for the same values of r for all four corner equilibria. If

$$a < b, g, \tag{40}$$

the stability of the corner equilibria depends on r. They are stable if $r > r_c$ and unstable if $r < r_c$, where in the symmetric viability model

$$r_c = a. \tag{41}$$

In this case a bifurcation analysis shows that there will be a curve of stable equilibria for $r < r_c$ approaching each corner equilibrium as r approaches r_c, if and only if

|b-g| > a.   (42)

If (42) holds and r is less than a, and r-a is sufficiently small, the symmetric viability model has four stable polymorphic equilibria.

Note, not surprisingly, that this complex behavior is occurs for values of the recombination parameter that are the same as the strength of selection. Numerical work confirms this. For example, let a=0.1, b=0.2, g=0.4. There are four stable polymorphic equilibria for r<0.1 and r sufficiently close to 0.1. Note also that in this case the symmetric equilibria (13), (14) are stable for 0<r<0.076. Not surprisingly, in the limit as r approaches 0.076 from above two stable unsymmetric equilibria approach each of the two symmetric equilibria. Another way of viewing this is to say that the unsymmetric equilibria arise from the symmetric equilibria as a result of a stable pitchfork bifurcation (Guckenheimer and Holmes, 1983). This same dependence on r holds for all other examples I have iterated and should hold whenever there are four stable polymorphic equilibria in the symmetric viability model.

General Two Locus Models

For fitnesses which do not fall into one of the special classes just listed, there are several possible directions in which to proceed to get partial results. One is to make assumptions about the fitnesses which allow the use of perturbation techniques. Another is to use various numerical approaches. I will discuss the kinds of results available from these in turn. I will begin with the analysis of models with weak selection, because this helps illustrate the presence of three regions in parameter space with different model behavior. Nagylaki (1976, 1977) (see also Hoppensteadt, 1976) used a clever perturbation argument to recover some of Fisher's fundamental theorem in this case.

I also have made use of another approach which is useful for investigating questions concerning behavior without making any *a priori* assumptions about the fitness matrix. The difficulty in even determining the equilibria of the model (10) lies in the fact that specifying the fitnesses and then solving for the equilibria is a nonlinear problem. However, if the equilibrium is specified, which is reasonable since it is much easier to measure allele or gametic frequencies than fitnesses, then solving for the equilibrium becomes a linear problem. Although the general solution is usually so complex as to be uninformative, this method can be used to examine numerically properties of large numbers of equilibria in an efficient fashion. Below, I will describe two results which emerged from this kind of study. Another method for obtaining useful information from the relationship between fitnesses and equilibria is to determine limits to this relationship using an optimization technique. Given the underlying linearity of the problem posed this way, by specifying the problem appropriately, linear programming methods can be used. I will describe the kinds of results possible with this approach as well.

Using the relationship between fitnesses and equilibria I generated large numbers of random equilibria and looked for two kinds of unusual behavior in two locus models. To search for stable cycles, I looked for equilibria with complex eigenvalues close to one in modulus, which might then lead to a Hopf bifurcation (Guckenheimer and Holmes, 1983) as a parameter (fitness or recombination) was varied. In fact several examples of stable cycling in two locus models were found by this method (Hastings, 1981a). Also, independently, stable cycles were found in the continuous time version (which involves some additional approximations) of the two locus two allele model by Akin (1982). These results show how different multiple locus dynamics can be, because the role of fitness as a Liapunov function (and the gradient structure of the continuous time version) in the one locus model precludes the possibility of stable cycles.

I also used this method of generating equilibria to search for counterexamples to the induced overdominance principle (see Ewens, 1979). This principle stated that at a stable equilibrium of a two (or multiple) locus model the induced fitnesses defined above must satisfy the conditions for stability at the induced system. For a one locus system these stability conditions are given in (6). Again several counterexamples were found which exhibited marginal underdominance (Hastings, 1981b). I was then able to examine several cases analytically using bifurcation techniques (Hastings, 1982).

Finally, as noted above information can be obtained about the equilibrium-fitness pairs by using optimization techniques, in particular linear programming

(Hastings, 1981c,1984,1986b). I used this approach to investigate several questions, in particular determining the limits to the relationship among recombination, strength of selection and disequilibrium at an equilibrium (Hastings, 1981) and the limits to the relationship among recombination, strength of epistasis and disequilibrium at an equilibrium (Hastings, 1986b).

The relationship among recombination, strength of selection and disequilibrium at an equilibrium was complex, but an examination of the numerical output revealed that the following bound always held. Denote by s the maximum strength of selection, the absolute value of the deviation of any fitness from that of the double heterozygote, which is assumed to be one. Then,

$$r|D| < s/10. \qquad\qquad (43)$$

This shows that strong selection is required to generate even moderate levels of disequilibrium.

I used a similar linear programming approach to answer the question -- how large must epistasis be to lead to a given level of disequilibrium ? Epistasis was measured as the maximum deviation (in absolute value) of the fitnesses of the four homozygotes from an additive model. Two results emerged. First, again strong epistasis is required to generate large levels of disequilibrium. Second, the greatest level of disequilibrium (for allele frequencies of one half at each locus) for a given strength of epistasis was achieved by the symmetric model. Thus, the intuition gained from the study of the symmetric model described above provides a good guide, a bound, to the role of epistasis in generating disequilibrium.

5. MULTILOCUS RESULTS

Among the first studies of more than two loci were simulations performed by Lewontin (1964a,b) and Franklin and Lewontin (1970). In the former case, a symmetric model was studied and linkage disequilibrium was shown to be important to a greater extent than would have been predicted by two locus models.

In the simulations of Franklin and Lewontin (1970) a symmetric, overdominant model with a large number of loci was studied. Here, they found very high disequilibrium values of all even orders, and for larger recombination rates than would have been predicted from two locus theory. For weaker and probably more realistic levels of selection, Clegg (1978) showed that this 'crystallization' effect was not important. Turelli and Ginzburg (1983) simulated a large number of 'random' fitness matrices and found that in general the intuition from one-locus two-allele models that heterosis is required for a stable equilibrium held for multilocus models.

Analytical work for more than two loci has followed the patterns described above for two locus results. If there is no recombination, the model simplifies to a one locus model, which has simple behavior, as noted above. Then perturbation techniques can be used to analyze the behavior for tightly linked systems, the case of small recombination rate (see Karlin and McGregor, 1972 and Karlin, 1978). The results obtained by the method of small recombination give insight into solutions with large levels of disequilibrium. Work remains to be done in considering the dependence of these equilibria on recombination parameters. Also, it is unknown what relationships among selection and recombination that would make these results based on small recombination good predictors of dynamical or equilibrium behavior, other than the fact that these results are valid for tight recombination, .

There has been work basically extending the approach taken by Nagylaki (1976, 1977) for understanding weak selection in two locus models to arbitrary numbers of loci by Shashahani (1979) and Akin (1979). Shashahani and Akin use approaches based on showing that recombination alone is a gradient system, and that with weak enough selection, the system approaches a state where the loci are nearly in linkage equilibrium--deviations from random combination of alleles at different loci are small. As in the tight linkage approach discussed above, it is not known, other than the fact that these results are valid for weak selection, when these results are good predictors of dynamical or equilibrium behavior.

There have been two approaches to studying the important cases which lie between very weak or very strong selection. One has been to explicitly examine three locus models in detail. Some of these are modifier models, with only two loci undergoing direct viability selection. The symmetric three locus model was studied by Feldman, Franklin and Thomson (1974), who explicitly found a large number of equilibrium solutions for weak recombination.

Another approach to the study of multilocus models has been to determine conditions under which an equilibrium with zero disequilibrium of any order, the product or Hardy-Weinberg equilibrium, is stable. This work is based on extensions of the nonepistatic models described above for two loci. For a wide variety of models conditions on local stability of the product equilibrium are obtained in Karlin and Liberman (1979a,1979b), Karlin and Avni (1981) and Karlin and Liberman (1982). An important result they obtain is that generally, for nonepistatic or symmetric fitnesses, if the product equilibrium is stable for a given recombination pattern, it is also stable for any pattern with 'more' recombination. In Karlin and Liberman (1978, 1979a, 1979b), it is shown that the product equilibrium is locally stable for the multilocus multiallele additive model as long as there is some recombination between all the loci. Kun and Ljubic (1979) prove global stability for the multilocus multiallele additive model through the use of two different Liapunov functions. Karlin and Liberman (1982) determine when the stability of the product equilibrium for a multilocus multiplicative model is controlled by two locus conditions.

There have also been several studies which have considered properties of equilibria without considering the role of stability. Ewens and Thomson (1977) derived a number of properties of marginal subsystems which are used below. Hastings (1984) showed that conditions for maintaining disequilibrium appear to be less stringent in three locus models than in two locus models.

Additive models for multiple loci, and nonepistatic models in general, are a natural class of multilocus models which have been extensively studied (see Karlin and Liberman, 1979a, 1979b, 1982). Nonepistatic models are models where the fitness of an individual depends on independent effects at different loci, and consequently has a simple representation in terms of the notation described above. In the additive model, the fitness of an individual is determined by summing effects at single loci, in the multiplicative model by multiplying effects at single loci. In this case, for example, if M_k is a matrix describing the fitness at each locus k, the matrix W describing fitnesses for the full system is given by:

$$W = M_1 \otimes M_2 \otimes ... \otimes M_n, \qquad (44)$$

as in equation (23). More general nonepistatic models are also possible (Karlin and Liberman, 1979a).

For all nonepistatic models there is a product equilibrium, an equilibrium where there is no correlation between the alleles at different loci (Karlin and Liberman, 1979a). Then, at this equilibrium, the frequency, x^*, of any gamete with alleles which have frequency x_i at locus i is given by:

$$x^* = x_1 \otimes x_2 \otimes ... \otimes x_n. \qquad (45)$$

Gametes have a frequency equal to the product of the frequencies that alleles at single loci would have, as determined by the single locus effects entering into the fitnesses. For the additive model, Karlin and Liberman (1978,1979a,1979b) have shown that this product equilibrium is locally stable independent of the number of loci or alleles involved, for positive recombination rates. Karlin and Liberman have also shown that this equilibrium is stable for the multiplicative model, if recombination rates are large enough.

I have performed some perturbation analyses on these equilibria to determine the effects of weak epistasis. A natural way to start the study of multilocus models is to assume weak additive epistasis. Experimental results reviewed in Simmons and Crow (1977) indicate that epistasis is nonzero, but weak. I considered models where the fitnesses deviate from those of a nonepistatic model, usually the additive model, by terms whose size is measured by a parameter δ. Thus δ is a measure of the size of the epistatic terms. Since I will restrict attention to those cases where the product equilibrium is locally stable, if epistasis is weak, δ is small, there remains a stable equilibrium for the model, close to the product equilibrium (see Karlin and MacGregor, 1972). The new equilibrium is a function of δ. As mentioned above, the equilibrium is known in the additive case, and each fitness in the case studied here is assumed to be 'close' to the additive model. Hence write all the fitnesses as:

$$w_{ij} = w_{ij;0} + \delta w_{ij;1} \tag{46}$$

where δ is a small parameter and $dw_{ij;1}$ gives the deviation away from the additive fitness, $w_{ij;0}$.

This perturbed equilibrium can be characterized by calculating the effect of the nonepistatic terms on mean fitness, allele frequencies and disequilibrium, up to first order in δ. The results were as follows:

RESULT 1. The change in the mean fitness due to weak epistasis does not depend on the recombination pattern, to a first approximation. As in the additive model, the mean fitness is independent of recombination, to a first approximation. This result does not depend on the additive model, but requires there to be a stable equilibrium with no disequilibrium, which is independent of the recombination rates.

RESULT 2. For a one locus marginal system fitnesses depend only on epistatic parameters actually involving the particular locus being considered.

RESULT 3. To lowest order, weak epistasis affects allele frequencies only through epistasis directly involving the particular locus in question. Moreover, only the epistasis is involved, and the recombination pattern does not enter.

The next results depend on the original nonepistatic model being additive.

RESULT 4. Pairwise disequilibria reflect additive epistasis involving only the loci being considered. The disequilibria vary as one divided by the probability of recombination rate between the loci.

RESULT 5. Three way disequilibria directly reflect the presence of additive epistasis at the loci involved. Three way disequilibria probably are smaller than pairwise disequilibria, since in most reasonable models the strength of epistasis goes down and also because there is an additional factor involving allele frequencies at the third locus. Finally, the dependence of disequilibrium on recombination is simple -- disequilibrium is roughly proportional to one divided by the probability that there is some recombination among the loci involved.

The following results was conjectured in Hastings (1986a), and shown to be true in a slightly different context by Barton (1986).
RESULT 6. For weak selection RESULT 5 is approximately true for higher order disequilibria with the appropriate changes in the wording.

These results all imply that the nonepistatic models provide a good guide to the behavior of multilocus systems with weak epistasis. Thus, it is an experimental challenge to determine how well natural systems with selection approximately the strength of recombination can be described by weakly epistatic models.

6. CONCLUSIONS

Mathematical research in multilocus population genetics has proceeded along different lines -- both in trying to determine to what extent the simple behavior predicted in the simple single locus models of population genetics is preserved, and to what extent more complex behavior is possible. A guide to the kinds of behavior possible is the strength of selection relative to recombination.

Even within the context of two locus models three different patterns are possible. For weak selection, behavior is simple because the system behaves as though different loci are independent. For strong selection, the system is simple, because for weak recombination, the system behaves much like a one locus system. For intermediate values of selection, complex behavior in terms of the number of equilibria and the conditions for stable equilibria occurs, and even stable cycles are

However, there are also selection patterns which lead to simple behavior for all combinations of selection and recombination strengths, even for multiple locus models. The results here suggest two challenges for the future. One is to determine what are the complete ranges of behavior possible for multiple locus models with arbitrary patterns of recombination and selection. The second is to determine what is the range of strengths of recombination and selection for which these complex behaviors are possible. Within the context of two locus models, some information is known, but the effects of adding more loci are less well understood.

BIBLIOGRAPHY

Akin, E., 1979 The Geometry of Population Genetics, Springer-Verlag, New York

Akin, E. 1982 Cycling in simple genetic systems. J. Math. Biology 13: 305-324.

Barton, N.H., 1986 The effects of linkage and density-dependent regulation on gene flow. Heredity 57:415-426.

Bodmer, W.F. and J. Felsenstein, 1967 Linkage and selection: Theoretical analysis of the deterministic two locus random mating model. Genetics 57:237-265.

Brown, A.H.D., 1975 Sample sizes required to detect linkage disequilibrium between two or three loci. Theor. Pop. Biol. 8:184-201.

Clegg, M.T., 1978 Dynamics of correlated genetic systems. II. Simulation studies of chromosomal segments under selection. Theor. Pop. Biol. 13:1 23.

Ewens, W.J. 1968 A genetic model having complex linkage behavior. Theoret. Appl. Genet. 38:140-143.

Ewens, W.J. 1969 Mean fitness increases when fitnesses are additive. Nature

Ewens, W., 1979 Mathematical Population Genetics. Springer-Verlag, New York.

Ewens, W. and G. Thomson, 1977 Properties of equilibria in multi-locus genetic systems. Genetics 87:807-819.

Falconer, D. Introduction to Quantitative Genetics, Second Edition, Longman,

Feldman, M.W., I. Franklin and G. Thomson, 1974 Selection in complex genetic systems I. The symmetric equilibria of the three-locus symmetric viability model. Genetics 76:135-162.

Feldman, M.W., 1986 An evolutionary reduction principle for genetic modifiers. Proc. Natl. Acad. Sci. USA 83:4824-4827.

Fisher, R.A., 1918 The correlation between relatives on the supposition of Mendelian inheritance. Trans. Roy. Soc. Edin. 52:399-433.

Franklin, I. and R. Lewontin, 1970 Is the gene the unit of selection? Genetics

Guckenheimer, J. and P. Holmes, 1983 Nonlinear Oscillations, Dynamical Systems, and Bifurcations of Vector Fields. Springer-Verlag, New York.

Hartl, D.L., 1980 Principles of Population Genetics. Sinauer and Associates, Sunderland, MA.

Hastings, Alan. 1981a. Stable cycling in discrete time genetic models. Proceedings of the National Academy of Sciences 78:7224-7225.

Hastings, Alan. 1981b. Marginal underdominance at a stable equilibrium. Proceedings of the National Academy of Sciences 78:6558-6559.

Hastings, Alan. 1981c. Disequilibrium, selection and recombination: limits in two-locus two-allele models. Genetics 98:659-668.

Hastings, Alan. 1982. Unexpected behavior in two locus genetic models: an analysis of marginal underdominance. Genetics 102:129-132

Hastings, Alan. 1984. Linkage disequilibrium, selection and recombination at three loci. Genetics 106:153-164.

Hastings, Alan. 1985. Four simultaneously stable polymorphic equilibria in two-locus two-allele models. Genetics 109:255-261.

Hastings, Alan. 1986a. Multilocus population genetics with weak epistasis. II. Equilibrium properties of multilocus models: What is the unit of selection? Genetics. 112:157-171.

Hastings, Alan. 1986b. Limits to the relationship among recombination, disequilibrium, and epistasis in two locus model. Genetics. 113:177-185.

Hedrick, P.W., S. Jain, and L. Holden, 1978 Multilocus systems in evolution. pp. 104-184. In: Evolutionary Biology, Edited by M.K. Hecht, W.C. Steere, and B. Wallace, Vol. 11, Plenum Press, New York.

Hoppensteadt, F., 1976 A slow selection analysis of two locus, two allele traits. Theor. Pop. Biol. 9:68-81

Karlin, S., 1975 General two-locus selection models: some objectives, results and interpretations. Theor. Pop. Biol. 7:364-398.

Karlin, S., 1978 Theoretical aspects of multi-locus selection balance I. pp. 503-587. In: Studies in Mathematical Biology Part II: Populations and Communities. Edited by S.A. Levin. Math. Assoc. Amer., Washington.

Karlin, S. and H. Avni, 1981 Analysis of central equilibria in multilocus systems: a generalized symmetric viability regime. Theor. Pop. Biol. 20:241-280.

Karlin, S. and M. Feldman, 1970a. Convergence to equilibrium of the two locus additive viability model. J. Appl. Probability 7:262-271.

Karlin, S. and M.W. Feldman, 1970 Linkage and selection -- two locus symmetric viability model. Theor. Pop. Biol. 1:39-71.

Karlin, S. and U. Liberman, 1978 The two-locus multi-allele additive viability model. J. Math. Biol. 5:201-211.

Karlin, S. and U. Liberman, 1979a Representation of nonepistatic selection models and analysis of multilocus Hardy-Weinberg equilibrium configurations. J. Math. Biol. 7:353-374.

Karlin, S. and U. Liberman, 1979b Central equilibria in multilocus systems. I. Generalized nonepistatic regimes. Genetics 91:777-798.

Karlin, S. and U. Liberman, 1982 The reduction property for central polymorphisms in nonepistatic systems. Theor. Pop. Biol. 22:69-95.

Karlin, S. and J. McGregor, 1972 Application of method of small parameters to multi-niche population genetic models. Theor. Pop. Biol. 3:186-209.

Kimura, M., 1983 The Neutral Theory of Molecular Evolution, Cambridge University Press, Cambridge.

Kingman, J.F.C., 1961 A mathematical problem in population genetics. Proc. Camb. Phil. Soc. 57:574-582

Kun, L.A. and Ju. I. Ljubic, 1979 Convergence to equilibrium under the action of additive selection in a multilocus multiallelic population. Soviet Math. Dokl. 20: 1380-1382.

Lewontin, R.C., 1964a The interaction of selection and linkage. I. General considerations; heterotic models. Genetics 49:49-67.

Lewontin, R.C., 1964b The interaction of selection and linkage. II. Optimum models. Genetics 50:757-782.

Lewontin, R.C., 1974 The Genetic Basis of Evolutionary Change. Columbia University Press, New York.

Lewontin, R.C. and K. Kojima, 1960 The evolutionary dynamics of complex polymorphisms. Evolution 14:458-472.

Nagylaki, T., 1976 The evolution of one- and two-locus systems. Genetics 83:583-

Nagylaki, T., 1977 The evolution of one- and two-locus systems. II. Genetics

85:347-354.

Nei, M., 1987 <u>Molecular Evolutionary Genetics</u>. Columbia University Press,

Provine, W.B., 1971 <u>The Origins of Theoretical Population Genetics</u>. University of Chicago Press, Chicago.

Shashahani, S. 1979 A new mathematical framework for the study of linkage and selection. Memoirs Amer. Math. Soc. 211.

Simmons, M.J. and J.F. Crow, 1977 Mutations affecting fitness in Drosophila populations. Ann. Rev. Genet. 11:49-78.

Turelli, M., 1982 *Cis-trans* effects induced by linkage disequilibrium. Genetics 72:157-168

Turelli, M., 1984 Heritable genetic variation via mutation-selection balance: Lerch's zeta meets the abdominal bristle. Theor. Pop. Biol. 25:138-193.

Turelli, M. and L. Ginzburg, 1983 Should individual fitness increase with heterozygosity. Genetics 104:191-209

Wright, S., 1935 Evolution in populations in approximate equilibrium. J. Genetics 30:257-266.

Wright, S., 1969 <u>Evolution and the Genetics of Populations, Vol. 2. The Theory of Gene Frequencies</u>, Univ. of Chicago Press, Chicago.

Division of Environmental Studies and Department of Mathematics
University of California
Davis, CA 95616

Lectures on Mathematics in the Life Sciences
Volume 20, 1989

THE DIFFUSION MODEL FOR MIGRATION
AND SELECTION

Thomas Nagylaki[1]

ABSTRACT. The diffusion model for clines maintained by migration and selection is formulated. The monoecious, diploid population is subdivided into an array of panmictic colonies that exchange migrants. Generations are discrete and nonoverlapping; the analysis is restricted to a single multiallelic locus. The relation between juvenile and adult subpopulation numbers is very general and includes both soft and hard selection. In the diffusion approximation, a partial differential equation and boundary conditions that incorporate spatial (and temporal) variation in the carrying capacity and migration rate are derived for the gene frequencies. These results hold for both adult and juvenile migration. A Lyapunov functional is obtained in the time homogeneous, spatially unidimensional, diallelic case. In one spatial dimension, transition conditions that simultaneously take into account discontinuities in the carrying capacity and migration rate are established: the gene frequencies are continuous, but their partial derivatives are not, their ratio being a simple function of the carrying capacities and migrational variances on the two sides of the inhomogeneity. The partial derivatives of the gene frequencies are continuous across a geographical barrier, whereas the gene frequencies themselves have a discontinuity proportional to the partial derivative at the barrier, the constant of proportionality being a measure of the difficulty of crossing the barrier.

1980 *Mathematics Subject Classification* (1985 *Revision*). Primary 92A10, 92A12; Secondary 35K57, 60J60.

[1] Supported by National Science Foundation Grant BSR-8512844.

1. INTRODUCTION

Natural populations are frequently distributed in space, and spatial inhomogeneities are very common. Changes in terrain and vegetation sometimes cause variation in the population density and dispersal rate. Rivers, sea shores, mountain ranges, canyons, etc., are obvious geographical barriers, but there are other possibilities. For instance, members of a species may be "reluctant" to cross from an unforested part of the habitat into the forested one, and vice versa, with the same effect.

The model derived here makes possible the study of spatial patterns in gene frequencies maintained by migration and selection. If they are stable and nonuniform, such spatial patterns are called clines. One of the aims of this paper is to illustrate the interesting, challenging biological and mathematical problems that arise in the formulation of models in population genetics, before these models are analyzed.

The continuous model for migration and selection is more tractable and sometimes biologically more appropriate than the discrete one. Although the continuous model can be formulated directly (Nagylaki, 1975; Fife, 1979, Chs. 1–2), biological interpretations and extensions (e.g., the nature of population regulation and the incorporation of random genetic drift) are clearer in the discrete case. Therefore, we shall derive the continuous model by approximating the discrete one.

For analyses of the discrete model, consult Nagylaki (1977, Ch. 6), Moody (1979, 1981), Karlin (1982), and references therein. The continuous model has also been extensively investigated; see Peletier (1976, 1978), Nagylaki (1978), ten Eikelder (1979), Nagylaki and Moody (1980), Fife and Peletier (1981), Henry (1981, pp. 314–319), Pauwelussen (1981), Pauwelussen and Peletier (1981), Yanagida (1982), Alikakos (1983), van der Meer (1983), Keller (1984), and references therein. Downham and Shah (1976), Diekmann (1980), and Lui (1986) have examined a hybrid model that is discrete in time but continuous in space.

In Section 2, a partial differential equation and boundary conditions that incorporate spatial (and temporal) variation in the population density and dispersal rate are deduced for the gene frequencies. We obtain a Lyapunov functional in the time-homogeneous, spatially unidimensional, diallelic case. Steep spatial gradients are often reasonably idealized as discontinuities; these cases are the most tractable analytically, and their investigation yields qualitative and quantitative insight. In Section 3, we establish the transition conditions for discontinuities in the population density and dispersal rate and for a geographical barrier.

2. THE PARTIAL DIFFERENTIAL EQUATION AND BOUNDARY CONDITIONS

In Subsections 2.1 and 2.2, we shall deduce the partial differential equation and boundary conditions in detail for adult migration in one dimension, generalize to several dimensions, and show that the same results hold for juvenile migration. We shall derive a Lyapunov functional in Subsection 2.3.

2.1. The Partial Differential Equation

2.1a. Adult migration. Generations are discrete and nonoverlapping. The monoecious, diploid population is subdivided into a linear array of panmictic colonies that exchange adults independently of genotype. Organisms confined to a river, riverbank, seashore, mountain range, etc., occupy linear habitats. The unidimensional model applies also to populations in bidimensional habitats if only one coordinate, such as latitude, altitude on a mountain range, or distance from a river or seashore, matters. Selection acts solely through viability differences: we assume that all fertilities are the same. We neglect mutation and random genetic drift. The analysis is restricted to a single locus with alleles A_i, $i = 1, 2, \ldots, r$. Let $N_k(t)$ denote the number of zygotes in deme k $(= 0, \pm 1, \pm 2, \ldots)$ in generation t $(= 0, 1, 2, \ldots)$. We designate the frequency of A_i in zygotes in deme k in generation t by $p_{i,k}(t)$. We shall consistently use the subscripts h, i, and j for alleles and k, ℓ, and n for demes. We summarize our model in the life cycle below, in which the prime signifies the next generation.

$$\text{Zygotes} \xrightarrow{\text{selection}} \text{Adults} \xrightarrow{\text{migration}} \text{Adults} \xrightarrow{\text{regulation}} \text{Adults} \xrightarrow{\text{reproduction}} \text{Zygotes}$$
$$N_k, p_{i,k} \qquad N_k^*, p_{i,k}^* \qquad N_k^*, p_{i,k}' \qquad N_k', p_{i,k}' \qquad N_k', p_{i,k}'$$

Selection and migration generally change both the deme sizes and the gene frequencies; population regulation changes only the former. If population regulation acted also during selection, the model would be unaltered. By permitting it to act after migration, we make constant deme sizes possible. Reproduction merely returns the demes to Hardy-Weinberg proportions. Our model applies to gametic dispersion if we use gene frequencies after dispersion. Let

$$\mathbf{p}_k = (p_{1,k}, p_{2,k}, \ldots, p_{r,k}) \tag{2.1}$$

represent the vector of gene frequencies in deme k. If we write the viability of an $A_i A_j$ individual in deme k in generation t as $w_{ij,k}(\mathbf{p}_k, t)$, then the mean viability of individuals that carry A_i and of all individuals in deme k in generation t read, respectively,

$$w_{i,k} = \sum_j w_{ij,k} p_{j,k}, \qquad \overline{w}_k = \sum_{i,j} w_{ij,k} p_{i,k} p_{j,k}. \tag{2.2}$$

After selection, the gene frequencies are given by

$$p^*_{i,k} = p_{i,k} w_{i,k} / \overline{w}_k. \tag{2.3}$$

Let $m_{k\ell}(t)$ designate the probability that, in generation t, an individual in deme k after migration was born in deme ℓ. Hence, we have at once the gene frequencies

$$p'_{i,k} = \sum_\ell m_{k\ell} p^*_{i,\ell} \tag{2.4}$$

in the next generation. The backward migration matrix $(m_{k\ell})$ reflects spatial variation not only of the migration rates, but also of the subpopulation numbers. To separate these two sources of variation, we introduce the forward migration matrix $(\widetilde{m}_{k\ell})$. Let $\widetilde{m}_{k\ell}(t)$ represent the probability that, in generation t, an individual in deme k migrates to deme ℓ. Then (Malécot, 1948; Nagylaki, 1977, p. 131)

$$m_{k\ell} = N^*_\ell \widetilde{m}_{\ell k} \Big/ \sum_n N^*_n \widetilde{m}_{nk}. \tag{2.5}$$

Substituting (2.5) into (2.4) yields

$$\left(\sum_\ell N^*_\ell \widetilde{m}_{\ell k} \right) p'_{i,k} = \sum_\ell N^*_\ell \widetilde{m}_{\ell k} p^*_{i,\ell}, \tag{2.6}$$

which must be supplemented by (2.3). We assume that $w_{ij,k}(\mathbf{p}_k, t)$, $N_k(t)$, and $\widetilde{m}_{\ell k}(t)$ are given and N^*_k is specified as $N^*_k = N^*_k(N_k, \mathbf{p}_k)$ for every i, j, k, and ℓ. This completes the description of the discrete model for $p_{i,k}(t)$.

To obtain a more tractable model, we posit that migration and selection are both weak. We scale space and time according to

$$x = k\epsilon, \quad y = \ell\epsilon, \quad \tau = \lambda t. \tag{2.7}$$

Thus, in the new units, ϵ signifies the distance between adjacent colonies and λ corresponds to one generation. We let $\epsilon \to 0$ and $\lambda \to 0$ so that ϵ^2/λ remains fixed.

Suppose that the selection intensity is λ:

$$w_{ij,k}(\mathbf{p}_k, t) = 1 + O(\lambda) \tag{2.8}$$

as $\lambda \to 0$, uniformly in x. If all the $w_{ij,k}$ are multiplied by the same constant, the model is unaltered, so the choice of constant in (2.8) entails no loss of generality. Inserting (2.8) into (2.3), we find

$$p_{i,k}^* = p_{i,k} + \lambda s_{i,k}(\mathbf{p}_k, t) + o(\lambda) \tag{2.9}$$

as $\lambda \to 0$, uniformly in x. In terms of the scaled variables, we write

$$N_k(t) = \rho(x,\tau), \quad p_{i,k}(t) = P_i(x,\tau), \quad s_{i,k}(\mathbf{p}_k, t) = S_i(\mathbf{P}, x, \tau), \tag{2.10}$$

in which the scale of N_k does not affect the model, $\mathbf{P} = \mathbf{P}(x,\tau)$, and S_i is always a nonlinear function of $\mathbf{P}$. If ρ, P_i, or S_i depend on ϵ, we assume that as $\lambda \to 0$, they converge to limits uniformly in x; henceforth, ρ, P_i, and S_i denote these limits.

For migration, we require the diffusion hypotheses

$$\lim_{\lambda \to 0} \frac{\epsilon}{\lambda} \sum_{\ell:\, |\ell-k|<\theta/\epsilon} (\ell - k)\widetilde{m}_{k\ell} = M(x,\tau), \tag{2.11a}$$

$$\lim_{\lambda \to 0} \frac{\epsilon^2}{\lambda} \sum_{\ell:\, |\ell-k|<\theta/\epsilon} (\ell - k)^2 \widetilde{m}_{k\ell} = V(x,\tau), \tag{2.11b}$$

$$\lim_{\lambda \to 0} \frac{1}{\lambda} \sum_{\ell:\, |\ell-k|\geq\theta/\epsilon} \widetilde{m}_{k\ell} = 0 \tag{2.11c}$$

uniformly in x for every fixed $\theta > 0$. Clearly, M and V represent the mean and variance (which, in the limit, is identical to the mean square) of the migrational displacement per new time unit in the new length units; the corresponding quantities in generations are λM and λV, respectively. We assume also that the partial derivatives of $\mathbf{P}$, ρ, M, and V that appear below are the uniform limits as $\lambda \to 0$ of the corresponding discrete quantities and are continuous.

Finally, we must relate N_k^* to N_k. In view of (2.8), with essentially no loss of biological generality, we may suppose

$$N_k^* = N_k[1 + O(\lambda)] \tag{2.12}$$

as $\lambda \to 0$, uniformly in x. Clearly, (2.12) includes both soft ($N_k^* = N_k$) and hard ($N_k^* = N_k \overline{w}_k$) selection (cf. Nagylaki, 1977, p. 132).

To deduce the partial differential equation, we multiply (2.6) by a suitably chosen test function, approximate using Taylor's Theorem and (2.9) to (2.12), pass from the sum to an integral, and equate the integrand to zero, as suggested by the derivation of the Kolmogorov Forward Equation (Gnedenko, 1962, pp. 363–366). For a bounded function $f_k(t) = F(x, \tau)$ such that F_{xx} is continuous, (2.7) and (2.11) lead to (cf. Feller, 1971, pp. 333–335)

$$\sum_k \widetilde{m}_{\ell k} f_k(t) = F(y, \tau) + \lambda M(y, \tau) F_y(y, \tau) + \tfrac{1}{2}\lambda V(y, \tau) F_{yy}(y, \tau) + o(\lambda) \tag{2.13}$$

as $\lambda \to 0$, in which the subscripts on the right signify partial derivatives. We use a nonnegative test function $\phi_k = \Phi(x)$ such that Φ_{xx} is continuous and that $\Phi(x)$, $\Phi_x(x)$, and $\Phi_{xx}(x)$ vanish unless $x_1 < x < x_2$. Multiply (2.6) by ϕ_k and sum over k:

$$\sum_{k,\ell} N_\ell^* \widetilde{m}_{\ell k} p'_{i,k} \phi_k = \sum_{k,\ell} N_\ell^* \widetilde{m}_{\ell k} p^*_{i,\ell} \phi_k. \tag{2.14}$$

With the aid of (2.9), (2.13), (2.7), and (2.10), we obtain

$$\sum_\ell N_\ell^* \{ P_i(y, \tau)[\Phi(y) + \lambda M(y, \tau)\Phi_y(y) + \tfrac{1}{2}\lambda V(y, \tau)\Phi_{yy}(y)]$$
$$+ \lambda S_i(\mathbf{P}, y, \tau)\Phi(y) + o(\lambda)\} \tag{2.15}$$

for the right side of (2.14). For the left side, (2.7), (2.10), and (2.13) give

$$\sum_{k,\ell} N_\ell^* \widetilde{m}_{\ell k}[P_i(x, \tau) + \lambda P_{i,\tau}(x, \tau) + O(\lambda^2)]\Phi(x)$$
$$= \sum_\ell N_\ell^* \{ P_i(y, \tau)\Phi(y) + \lambda M(y, \tau[P_i(y, \tau)\Phi(y)]_y$$
$$+ \tfrac{1}{2}\lambda V(y, \tau)[P_i(y, \tau)\Phi(y)]_{yy} + \lambda P_{i,\tau}(y, \tau)\Phi(y) + o(\lambda)\}. \tag{2.16}$$

Equating (2.15) and (2.16), we get

$$\sum_{\ell} N_{\ell}^{*} \big\{ \big[M(y,\tau)P_{i,y}(y,\tau) + \tfrac{1}{2}V(y,\tau)P_{i,yy}(y,\tau) + P_{i,\tau}(y,\tau)$$
$$- S_i(\mathbf{P},y,\tau) \big] \Phi(y) + V(y,\tau)P_{i,y}(y,\tau)\Phi_y(y) + o(1) \big\} = 0. \quad (2.17)$$

As $\lambda \to 0$, (2.17) converges to an integral; (2.10) and (2.12) enable us to replace N_{ℓ}^{*} by $\rho(y,\tau)$. We integrate the last term in (2.17) by parts and recall the assumptions on Φ to derive

$$\int_{x_1}^{x_2} \big\{ \rho(y,\tau)[M(y,\tau)P_{i,y}(y,\tau) + \tfrac{1}{2}V(y,\tau)P_{i,yy}(y,\tau) + P_{i,\tau}(y,\tau)$$
$$- S_i(\mathbf{P},y,\tau)] - [\rho(y,\tau)V(y,\tau)P_{i,y}(y,\tau)]_y \big\} \Phi(y)dy = 0. \quad (2.18)$$

Since Φ is arbitrary, the brace must vanish. We put the resulting partial differential equation in the form (assuming $\rho(x,\tau) > 0$ for all x and τ)

$$P_{i,\tau} = \tfrac{1}{2}VP_{i,xx} + \big[\rho^{-1}(\rho V)_x - M\big]P_{i,x} + S_i. \quad (2.19)$$

We can express the migration terms in (2.19) in terms of

$$J = M\rho - \tfrac{1}{2}(V\rho)_x, \quad (2.20a)$$

the flux of individuals under migration, and

$$K_i = M\rho P_i - \tfrac{1}{2}(V\rho P_i)_x = JP_i - \tfrac{1}{2}V\rho P_{i,x}, \quad (2.20b)$$

the flux of A_i alleles under migration (Nagylaki, 1975):

$$P_{i,\tau} = \rho^{-1}(P_i J_x - K_{i,x}) + S_i. \quad (2.21)$$

If $V(x,\tau) > 0$ for all x and τ, we can conveniently rewrite (2.19) as (Fife, 1979, p. 62)

$$P_{i,\tau} = (2\rho^2 V)^{-1}\big[(\rho V)^2 P_{i,x}\big]_x - MP_{i,x} + S_i. \quad (2.22)$$

We can easily generalize (2.19), (2.20), and (2.21) to d spatial dimensions. Let k and ℓ represent d-dimensional multi-indices; x and y are now vectors in d dimensions. Then (2.11c) still holds, but (2.11a) and (2.11b) become

$$\lim_{\lambda \to 0} \frac{\epsilon}{\lambda} \sum_{\ell:\ |\ell-k|<\theta/\epsilon} (\ell-k)_\alpha \widetilde{m}_{k\ell} = M_\alpha(x,\tau), \qquad (2.23a)$$

$$\lim_{\lambda \to 0} \frac{\epsilon^2}{\lambda} \sum_{\ell:\ |\ell-k|<\theta/\epsilon} (\ell-k)_\alpha (\ell-k)_\beta \widetilde{m}_{k\ell} = V_{\alpha\beta}(x,\tau), \qquad (2.23b)$$

where $\alpha, \beta = 1, 2, \ldots, d$. The test function $\Phi(x)$ has continuous second-order partial derivatives, and $\Phi(x)$, $\Phi_{x_\alpha}(x)$, and $\Phi_{x_\alpha x_\beta}(x)$ vanish for every α and β unless $x_{1,\gamma} < x_\gamma < x_{2,\gamma}$ for every γ. Following the unidimensional derivation, but now employing the multidimensional Taylor Theorem, we deduce the generalization of (2.19):

$$P_{i,\tau} = \sum_{\alpha,\beta} [\tfrac{1}{2} V_{\alpha\beta} P_{i,x_\alpha x_\beta} + \rho^{-1}(\rho V_{\alpha\beta})_{x_\alpha} P_{i,x_\beta}] - \sum_\alpha M_\alpha P_{i,x_\alpha} + S_i. \qquad (2.24)$$

The multidimensional fluxes

$$J_\alpha = M_\alpha \rho - \tfrac{1}{2} \sum_\beta (V_{\alpha\beta}\rho)_{x_\beta}, \qquad (2.25a)$$

$$K_{i,\alpha} = M_\alpha \rho P_i - \tfrac{1}{2} \sum_\beta (V_{\alpha\beta}\rho P_i)_{x_\beta} = J_\alpha P_i - \tfrac{1}{2} \sum_\beta V_{\alpha\beta}\rho P_{i,x_\beta} \qquad (2.25b)$$

enable us to recast (2.24) into the multidimensional generalization of (2.21):

$$P_{i,\tau} = \rho^{-1} \sum_\alpha (P_i J_{\alpha,x_\alpha} - K_{i,\alpha,x_\alpha}) + S_i \qquad (2.26)$$

(Nagylaki, 1975). Notice that, as expected for a conservation law (with a source term) for numbers (not proportions) of alleles, only the divergence of the fluxes appears in (2.26).

2.1b. Juvenile migration.   Our formal scheme is now the following.

$$\begin{array}{ccccccccc} \text{Zygotes} & \xrightarrow{\text{migration}} & \text{Zygotes} & \xrightarrow{\text{selection}} & \text{Adults} & \xrightarrow{\text{regulation}} & \text{Adults} & \xrightarrow{\text{reproduction}} & \text{Zygotes} \\ N_k, p_{i,k} & & N_k^*, p_{i,k}^* & & N_k^{**}, p_{i,k}' & & N_k', p_{i,k}' & & N_k', p_{i,k}' \end{array}$$

After migration, the ordered genotypic frequencies are

$$q_{ij,k} = \sum_\ell m_{k\ell} p_{i,\ell} p_{j,\ell}. \qquad (2.27)$$

Instead of (2.5), we have

$$m_{k\ell} = N_\ell \widetilde{m}_{\ell k} \Big/ \sum_n N_n \widetilde{m}_{nk}. \tag{2.28}$$

Selection produces the allelic frequencies

$$p'_{i,k} = \sum_j q_{ij,k} w_{ij,k} \Big/ \sum_{h,j} q_{hj,k} w_{hj,k}. \tag{2.29}$$

Substituting (2.27) and (2.28) into (2.29) leads to our basic equation:

$$\left(\sum_{h,j,\ell} N_\ell w_{hj,k} p_{h,\ell} p_{j,\ell} \widetilde{m}_{\ell k} \right) p'_{i,k} = \sum_{j,\ell} N_\ell w_{ij,k} p_{i,\ell} p_{j,\ell} \widetilde{m}_{\ell k}. \tag{2.30}$$

Suppose

$$w_{ij,k} = 1 + \lambda u_{ij,k} + o(\lambda) \tag{2.31}$$

as $\lambda \to 0$, uniformly in x, and define

$$u_{i,k} = \sum_j u_{ij,k} p_{j,k}, \qquad \overline{u}_k = \sum_{i,j} u_{ij,k} p_{i,k} p_{j,k}, \tag{2.32}$$

$$s_{i,k} = p_{i,k}(u_{i,k} - \overline{u}_k). \tag{2.33}$$

Then we use (2.7), (2.10), (2.11), (2.31), (2.32), and (2.33) to approximate (2.30), following the method applied to (2.6). Although the discrete models (2.6) and (2.30) are quite different, this procedure again yields (2.18), and hence the same diffusion model (2.19). Generalizing to several dimensions, we can establish easily that (2.24) also holds for juvenile migration.

2.2. The Boundary Conditions

2.2a. Adult migration. In a continuous time, continuous-space model, $\rho(x, \tau)$ is not prescribed arbitrarily; rather, its evolution is determined by migration, selection, and population regulation. If the habitat is $[0, \infty)$, with no migration to the left of the origin, the flux of individuals and of A_i alleles must vanish at the origin:

$$J(0, \tau) = K_i(0, \tau) = 0. \tag{2.34}$$

From (2.20b) and (2.34) we infer the desired boundary condition,

$$P_{i,x}(0, \tau) = 0. \tag{2.35}$$

Such a Neumann condition was derived in the neutral case for nearest-neighbor migration by Fleming and Su (1974).

To prove (2.35) for our discrete model, consider colonies at $0, 1, 2, \ldots$ and assume that $m_{0\ell} = 0$ if $\ell > L$, for some fixed integer $L < \infty$. Inserting (2.9) into (2.4), we have

$$p'_{i,0} = \sum_{\ell=0}^{L} m_{0\ell} p_{i,\ell} + O(\lambda) \tag{2.36}$$

as $\lambda \to 0$, whence

$$\frac{\lambda}{\epsilon} \left(\frac{p'_{i,0} - p_{i,0}}{\lambda} \right) = \sum_{\ell=0}^{L} m_{0\ell} \left(\frac{p_{i,\ell} - p_{i,0}}{\epsilon} \right) + O(\epsilon). \tag{2.37}$$

Letting $\lambda \to 0$, we obtain

$$P_{i,x}(0, \tau) \sum_{\ell=0}^{L} \ell m_{0\ell} = 0, \tag{2.38}$$

which establishes (2.35).

In several dimensions, suppose the boundary passes through the origin and has normal vector ν there. Then

$$\sum_{\alpha} \nu_\alpha J_\alpha(0, \tau) = \sum_{\alpha} \nu_\alpha K_{i,\alpha}(0, \tau) = 0. \tag{2.39}$$

From (2.25b) and (2.39) we get

$$\sum_{\alpha,\beta} \nu_\alpha V_{\alpha\beta} P_{i,x_\beta}(0, \tau) = 0. \tag{2.40}$$

The boundary condition (2.40) was verified for the two-dimensional discrete model by choosing the x_1-axis to be orthogonal to the boundary and restricting migration so that $\widetilde{m}_{k\ell} = 0$ if $|\ell_1 - k_1| > 1$ or $|\ell_2 - k_2| > 1$.

Since it is numbers rather than frequencies of alleles that diffuse, therefore (2.35) and (2.40), unlike (2.34) and (2.39), are not zero-flux conditions.

2.2b. Juvenile migration. Substituting (2.31) into (2.29) and utilizing (2.27), we derive

$$p'_{i,k} = \sum_{\ell} m_{k\ell} p_{i,\ell} + O(\lambda) \tag{2.41}$$

as $\lambda \to 0$. Comparing (2.41) with (2.36) immediately demonstrates that (2.35) and (2.40) hold also for juvenile migration.

2.3. The Lyapunov Functional

We take the finite segment $[a, b]$ as the habitat, specialize to two alleles, and assume that the problem is time-homogeneous. Then (2.22) and (2.35) become

$$P_\tau = (2\rho^2 V)^{-1}[(\rho V)^2 P_x]_x - M P_x + S, \tag{2.42a}$$

$$P_x(a, \tau) = P_x(b, \tau) = 0, \tag{2.42b}$$

where

$$\rho = \rho(x) > 0, \quad M = M(x), \quad V = V(x) > 0, \quad S = S[P(x, \tau), x]. \tag{2.42c}$$

We define

$$T(P, x) = \int_0^P S(R, x) dR \tag{2.43}$$

and try the Lyapunov functional

$$I(\tau) = \int_a^b \{G(x)[P_x(x, \tau)]^2 - H(x)T[P(x, \tau), x]\} dx, \tag{2.44}$$

where G and H are to be determined.

Differentiating (2.44), using (2.43), integrating by parts, and invoking (2.42b), we find

$$\frac{dI}{d\tau}(\tau) = -\int_a^b H P_\tau[2H^{-1}(G P_x)_x + S] dx. \tag{2.45}$$

If we can rewrite (2.42a) as

$$P_\tau = 2H^{-1}(G P_x)_x + S, \tag{2.46}$$

with $H(x) > 0$, then (2.45) simplifies to

$$\frac{dI}{d\tau}(\tau) = -\int_a^b H(x)[P_\tau(x, \tau)]^2 dx \leq 0. \tag{2.47}$$

In this case, since $I(\tau)$ is bounded below, it is a Lyapunov functional, and its existence suggests that $P(x, \tau)$ converges as $\tau \to \infty$. Comparing (2.46) with

(2.42a) leads (up to an arbitrary constant multiplying both G and H) to

$$G(x) = \tfrac{1}{4}[\rho(x)V(x)]^2 \exp\left[-2\int^x \frac{M(y)}{V(y)}\,dy\right], \quad H(x) = \frac{4}{V(x)}\,G(x). \quad (2.48)$$

For isotropic migration, $M(x) = 0$, so (2.48) reduces to

$$G(x) = \tfrac{1}{4}[\rho(x)V(x)]^2, \quad H(x) = [\rho(x)]^2 V(x). \quad (2.49)$$

It is interesting to note that the exponential in (2.48) is the density of the scale function for the diffusion process generated by migration (cf. Karlin and Taylor, 1981, p. 194). Our Lyapunov functional generalizes those of Fleming (1975) and Barton (1979).

3. THE TRANSITION CONDITIONS

To derive the partial differential equation (2.19), we had to assume that the carrying capacity (ρ) and the mean and variance of the migrational displacement (M and V) are smooth. As discussed in Section 1, however, steep spatial gradients in these quantities are often reasonably idealized as discontinuities, and these cases yield considerable biological insight. Finite discontinuities in M have no dramatic effects (since P and P_x remain continuous), but discontinuities in ρ and V do. The examples in Section 1 suggest that ρ and V may often have discontinuities at the same location. We investigate this situation in Section 3.1; clearly, discontinuities in only one of ρ and V are included as special cases. In Section 3.2, we establish the transition conditions for a geographical barrier.

We consider only adult migration in one spatial dimension; the analyses of Section 2 indicate that the transition conditions would be the same for juvenile migration.

3.1. Discontinuity in the Carrying Capacity and Migration Rate

Suppose M is bounded and ρ and V are discontinuous at the origin. The transition conditions can be deduced from (2.22) by following Fife (1979, pp. 67–72). Since the diffusion hypotheses (2.11) are local, (2.22) holds if $x \neq 0$. We can use (2.22) everywhere if we interpret the discontinuities as limits of steep gradients. This enables us to derive transition conditions at the origin, so that we need (2.22) only for $x \neq 0$. Discontinuities in ρ and V produce no discontinuity in P_i:

$$P_i(0-,\tau) = P_i(0+,\tau). \qquad (3.1a)$$

To see that $P_{i,x}$ must be discontinuous at $x = 0$, multiply (2.22) by $2\rho^2 V$, integrate from $-x_0$ ($x_0 > 0$) to x_0, and let $x_0 \to 0$. We obtain

$$[\rho(0-,\tau)V(0-,\tau)]^2 P_{i,x}(0-,\tau) = [\rho(0+,\tau)V(0+,\tau)]^2 P_{i,x}(0+,\tau). \qquad (3.1b)$$

The dependence on ρ agrees with unpublished work of Sawyer (personal communication). The dependence on V, derived from a discrete model, however, was V instead of the V^2 in (3.1b) (Nagylaki, 1976). To show that this apparent discrepancy reflects the sensitive dependence of the transition conditions on the migration pattern near the discontinuity, consider first the discrete model displayed in Figure 1.

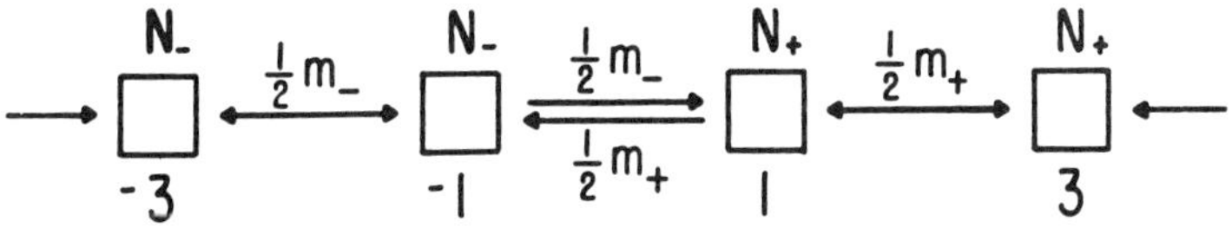

Figure 1. Isotropic migration with jumps in the carrying capacity and migration rate.

We scale space and time as in (2.7), but for convenience place colonies only at the odd integers. Let

$$N_k = \begin{cases} N_-, & k = -1, -3, \ldots, \\ N_+, & k = 1, 3, \ldots, \end{cases} \qquad (3.2)$$

$$\widetilde{m}_{k,k-2} = \widetilde{m}_{k,k+2} = \begin{cases} \frac{1}{2}m_-, & k = -1, -3, \ldots, \\ \frac{1}{2}m_+, & k = 1, 3, \ldots, \end{cases} \qquad (3.3a)$$

$$\widetilde{m}_{k\ell} = 0, \qquad |\ell - k| > 2; \qquad (3.3b)$$

$N_\pm$ and $m_\pm$ designate constants. Migration is obviously *isotropic*,

$$\widetilde{m}_{k,k-\ell} = \widetilde{m}_{k,k+\ell} \qquad (3.4)$$

for every odd k and even ℓ, and therefore (2.11a) implies that $M \equiv 0$. We again define $p_{i,k}(t) = P_i(x, \tau)$ and introduce the constants $\rho_\pm$ and $V_\pm$:

$$\rho_\pm = \tfrac{1}{2}N_\pm \quad V_\pm = m_\pm (2\epsilon)^2/\lambda, \tag{3.5}$$

which denote the population density and the migrational variance for $x \gtrless 0$.

We do *not* posit that $w_{ij,k} = w_{ij,\pm}$ for $k \gtrless 0$. This restrictive assumption would not facilitate the analysis because, despite (3.2), spatial variation in the gene frequencies may still preclude the simplification $N_k^* = N_\pm^*$, $k \gtrless 0$, as it clearly would for hard selection. Of course, for soft selection, (3.2) shows that $N_k^* = N_\pm$, $k \gtrless 0$.

Since $\tilde{m}_{k\ell} = \tilde{m}_{\ell k}$ for $|k| > 1$, therefore (2.5), (2.12), and (3.2) yield

$$m_{k\ell} = \tilde{m}_{k\ell} + O(\lambda), \quad |k| > 1, \tag{3.6a}$$

as $\lambda \to 0$. From (2.5), (2.12), (3.2), (3.3), and (3.5) we get

$$m_{\pm 1, \pm 3} = m_\pm \rho_\pm (2B_\pm)^{-1} + O(\lambda), \tag{3.6b}$$

$$m_{\pm 1, \mp 1} = m_\mp \rho_\mp (2B_\pm)^{-1} + O(\lambda), \tag{3.6c}$$

$$m_{\pm 1, \pm 1} = (1 - m_\pm)\rho_\pm B_\pm^{-1} + O(\lambda) \tag{3.6d}$$

as $\lambda \to 0$, where

$$B_\pm = (1 - \tfrac{1}{2}m_\pm)\rho_\pm + \tfrac{1}{2}m_\mp \rho_\mp. \tag{3.6e}$$

Substituting (2.9) and (3.6) into (2.4) gives

$$B_\pm(p'_{i,\pm 1} - p_{i,\pm 1}) = \tfrac{1}{2}m_\pm \rho_\pm(p_{i,\pm 3} - p_{i,\pm 1}) - \tfrac{1}{2}m_\mp \rho_\mp(p_{i,\pm 1} - p_{i,\mp 1}) + O(\lambda) \tag{3.7}$$

as $\lambda \to 0$, which we rearrange with the aid of (3.5) as

$$4\epsilon B_\pm\left(\frac{p'_{i,\pm 1} - p_{i,\pm 1}}{\lambda}\right) = \rho_\pm V_\pm\left(\frac{p_{i,\pm 3} - p_{i,\pm 1}}{2\epsilon}\right) \mp \rho_\mp V_\mp D_{i,\epsilon} + O(\epsilon), \tag{3.8}$$

where

$$D_{i,\epsilon} = (p_{i,1} - p_{i,-1})/(2\epsilon). \tag{3.9}$$

Now let $\lambda \to 0$. Both parentheses in (3.8) are discrete partial derivatives, and neither crosses the discontinuity. Therefore, by the assumption below (2.11), both have limits, and hence so does $\rho_{\mp} V_{\mp} D_{i,\epsilon}$, which implies that

$$D_{i,0} = \lim_{\epsilon \to 0} (p_{i,1} - p_{i,-1})/(2\epsilon) \tag{3.10}$$

exists. Equation (3.10) confirms (3.1a). In the limit, (3.8) informs us

$$\pm \rho_{\pm} V_{\pm} P_{i,x}(0\pm, \tau) \mp \rho_{\mp} V_{\mp} D_{i,0} = 0. \tag{3.11}$$

Eliminating $D_{i,0}$ from the two equations contained in (3.11), we conclude

$$(\rho_- V_-)^2 P_{i,x}(0-, \tau) = (\rho_+ V_+)^2 P_{i,x}(0+, \tau), \tag{3.12}$$

in agreement with (3.1b).

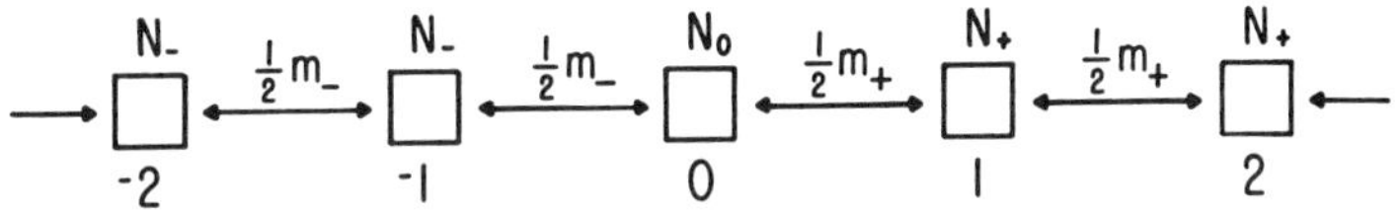

Figure 2. Symmetric migration with jumps in the carrying capacity and migration rate.

Next, consider the migration pattern in Figure 2, which is the same as that in Figure 1 of Nagylaki (1976), but here we incorporate also a possible discontinuity in the population density. Whereas the migration pattern in Figure 1 is isotropic (as defined in (3.4)), the pattern shown in Figure 2 is *symmetric*:

$$\widetilde{m}_{k\ell} = \widetilde{m}_{\ell k} \tag{3.13}$$

for every k and ℓ. The important point is that if $m_+ \neq m_-$, then $M \neq 0$ at the origin. An analysis paralleling the one just presented establishes that (3.1a) still holds, but now (3.12) must be replaced by

$$\rho_-^2 V_- P_{i,x}(0-, \tau) = \rho_+^2 V_+ P_{i,x}(0+, \tau), \tag{3.14}$$

in agreement with Nagylaki (1976).

How do we reconcile (3.14) with our derivation of (3.1b) directly from the partial differential equation (2.22)? Fife (personal communication) has proved that for symmetric migration

$$M(x,\tau) = \tfrac{1}{2}V_x(x,\tau). \tag{3.15}$$

To see this, first use (3.13) in (2.6) to obtain

$$\left(\sum_\ell \tilde{m}_{k\ell} N_\ell^*\right) p'_{i,k} = \sum_\ell \tilde{m}_{k\ell} N_\ell^* p_{i,\ell}^*, \tag{3.16}$$

then invoke (2.13) to approximate both sides of (3.16), with the result

$$P_{i,\tau} = (2\rho^2)^{-1}V(\rho^2 P_{i,x})_x + M P_{i,x} + S_i. \tag{3.17}$$

Equating the right sides of (2.22) and (3.17) establishes (3.15).

From (3.15) we see that if migration is symmetric and V has a discontinuity, then M contains a delta function, and hence the derivation of (3.1b) fails. In fact, (3.15) simplifies both (2.22) and (3.17) to

$$P_{i,\tau} = (2\rho^2)^{-1}(\rho^2 V P_{i,x})_x + S_i. \tag{3.18}$$

The argument leading to (3.1b) now reveals

$$[\rho(0-,\tau)]^2 V(0-,\tau)P_{i,x}(0-,\tau) = [\rho(0+,\tau)]^2 V(0+,\tau)P_{i,x}(0+,\tau), \tag{3.19}$$

in agreement with (3.14).

The choice between isotropy (3.4) and symmetry (3.13) is a biological one: isotropy would result from random choice of direction, whereas symmetry might occur if the difficulty of traversing a variable terrain were the main obstacle to migration. Fortunately, the difference is quantitative, not qualitative. It is, however, possible that migration schemes different from those in Figures 1 and 2 would yield transition conditions different from (3.1b) and (3.19).

3.2. Migration Across a Geographical Barrier

Slatkin (1973) was the first to study this problem; the model in Figure 3 was employed in Nagylaki (1976).

We retain the scaling (2.7), but place colonies only at the odd integers. Let $N_k = N$,

$$\tilde{m}_{k,k+2} = \tfrac{1}{2}m, \quad k = 1, \pm3, \pm5, \ldots, \tag{3.20a}$$

$$\tilde{m}_{-1,1} = \tfrac{1}{2}cm, \tag{3.20b}$$

$$\tilde{m}_{k\ell} = \tilde{m}_{\ell k}, \quad k, \ell = \pm1, \pm3, \ldots, \tag{3.20c}$$

$$\tilde{m}_{k\ell} = 0, \quad |\ell - k| > 2; \tag{3.20d}$$

N, m, and c represent constants. Migration across the symmetric geographical barrier located at the origin is decreased by a factor c $(0 \le c \le 1)$. We define $p_{i,k}(t) = P_i(x, \tau)$ and introduce the constants

$$\rho = \tfrac{1}{2}N, \quad V = m(2\epsilon)^2/\lambda, \tag{3.21}$$

which signify the population density and the migrational variance.

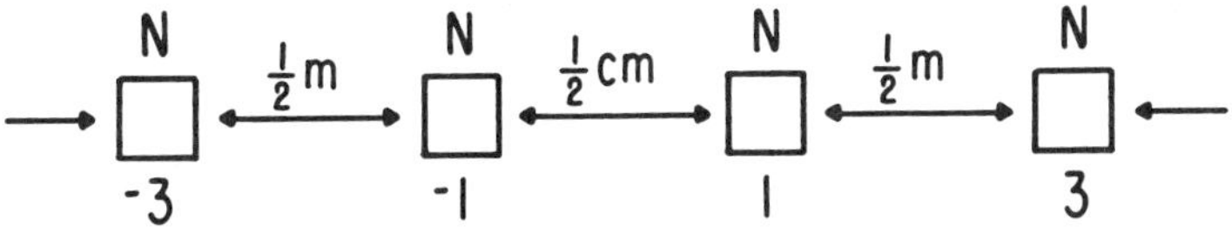

Figure 3. Migration across a geographical barrier.

Since $N_k = N$, from (2.5), (2.12), and (3.20c) we see immediately

$$m_{k\ell} = \tilde{m}_{k\ell} + O(\lambda) \tag{3.22}$$

as $\lambda \to 0$. Thus, symmetry of the forward migration matrix implies approximate symmetry of the backward migration matrix. Appealing to (2.4), (2.9), (3.22), (3.20), and (3.21), we find

$$4\epsilon\left(\frac{p'_{i,\pm1} - p_{i,\pm1}}{\lambda}\right) = V\left(\frac{p_{i,\pm3} - p_{i,\pm1}}{2\epsilon}\right) \mp \left(\frac{c}{2\epsilon}\right)V\Delta_{i,\epsilon} + O(\epsilon) \tag{3.23}$$

as $\lambda \to 0$, where

$$\Delta_{i,\epsilon} = p_{i,1} - p_{i,-1}. \tag{3.24}$$

Now let $\lambda \to 0$. Both parentheses in (3.23) are discrete partial derivatives, and neither crosses the discontinuity. Therefore, by the assumption below (2.11), both have limits. If $c/(2\epsilon) \to \infty$ (as would happen if c were positive and independent of ϵ), then $\Delta_{i,\epsilon} \to 0$. In this case, $P_i(x,\tau)$ is continuous and the barrier disappears in the diffusion limit. Therefore, we assume $c/(2\epsilon) \to \gamma < \infty$. Then (3.23) becomes (Nagylaki, 1976)

$$P_{i,x}(0\pm,\tau) = \gamma[P_i(0+,\tau) - P_i(0-,\tau)]. \tag{3.25}$$

So $P_{i,x}$ is continuous, but P_i has a discontinuity at the origin proportional to the slope there.

If $\gamma = 0$, there is no migration across the barrier in the diffusion limit, and (3.25) confirms (2.35).

Sawyer (personal communication) has given an illuminating interpretation of the transmission parameter γ: The probability that an immortal individual at x (> 0) crosses the barrier before reaching y ($> x$) is

$$Q(x) = (y - x)/(y + \gamma^{-1}). \tag{3.26}$$

As expected intuitively, Q is an increasing function of γ. Observe that the probability than an individual at the barrier crosses it before reaching the point γ^{-1} is $1/2$. Biologically, an "immortal individual" refers to a lineage that survives until it either crosses the barrier or reaches y. Thus, (3.26) may allow the estimation of γ from the migration pattern without estimating c.

To prove (3.26), let q_k denote the probability that an immortal individual at k (> 0) crosses the barrier before reaching ℓ ($> k$). A glance at Figure 3 informs us

$$q_k = \tfrac{1}{2}mq_{k-2} + (1 - m)q_k + \tfrac{1}{2}mq_{k+2}, \qquad 1 < k < \ell, \tag{3.27a}$$

$$q_1 = \tfrac{1}{2}cm + \left[1 - \tfrac{1}{2}(1 + c)m\right]q_1 + \tfrac{1}{2}mq_3, \tag{3.27b}$$

$$q_\ell = 0. \tag{3.27c}$$

The general solution of (3.27a) reads

$$q_k = b_1 - b_2 k, \tag{3.28}$$

where b_1 and b_2 designate constants. Enforcing (3.27b) and (3.27c) leads to

$$q_k = c(\ell - k)/[c(\ell - 1) + 2], \tag{3.29}$$

the diffusion limit of which is (3.26). Alternatively, we can deduce (3.26) directly from the diffusion limit of (3.27).

ACKNOWLEDGMENT

I am very grateful to Professor Paul Fife for extensive, very helpful correspondence. I thank Professor Stanley Sawyer for many penetrating observations and for communicating his unpublished results to me.

REFERENCES

Alikakos, N. D., 1983 Quantitative maximum principles and strongly coupled gradient-like reaction-diffusion systems. Proc. Roy. Soc. Edinb. *94A*: 265–280.

Barton, N. H., 1979 The dynamics of hybrid zones. Heredity *43*: 291–309.

Diekmann, O., 1980 Clines in a discrete time model in population genetics. In *Biological Growth and Spread* (ed. W. Jäger, H. Rost, and P. Tautu). Pp. 267–278. Springer, Berlin.

Downham, D. Y. and S. M. M. Shah, 1976 A sufficiency condition for the stability of an equilibrium. Adv. Appl. Prob. *8*: 4–7.

Eikelder, H. M. M., 1979 A non-linear diffusion problem arising in population genetics. Report NA-25, Delft University of Technology.

Feller, W., 1971 *An Introduction to Probability Theory and Its Applications*. Vol. II, 2nd ed. Wiley, New York.

Fife, P. C., 1979 *Mathematical Aspects of Reacting and Diffusing Systems*. Springer, Berlin.

Fife, P. C. and L. A. Peletier, 1981 Clines induced by variable selection and migration. Proc. Roy. Soc. Lond. B *214*: 99–123.

Fleming, W. H., 1975 A selection-migration model in population genetics. J. Math. Biol. 2: 219–233.

Fleming, W. H. and C.-H. Su, 1974 Some one-dimensional migration models in population genetics theory. Theor. Pop. Biol. *5*: 531–549.

Gnedenko, B. V., 1962 *The Theory of Probability*. 4th ed. Chelsea, New York.

Henry, D., 1981 *Geometric Theory of Semilinear Parabolic Equations*. Springer, Berlin.

Karlin, S., 1982 Classifications of selection-migration structures and conditions for a protected polymorphism. Evol. Biol. *14*: 61–204.

Karlin, S. and H. M. Taylor, 1981 *A Second Course in Stochastic Processes*. Academic Press, New York.

Keller, J. B., 1984 Genetic variability due to geographical inhomogeneity. J. Math. Biol. *20*: 223–230.

Lui, R., 1986 A nonlinear integral operator arising from a model in population genetics. IV. Clines. SIAM J. Math. Anal. *17*: 152–168.

Malécot, G., 1948 *Les mathématiques de l'hérédité*. Masson, Paris. Extended translation: *The Mathematics of Heredity*. Freeman, San Francisco (1969).

van der Meer, J. J. E., 1983 Clines induced by a geographical barrier. Preprint.

Moody, M. E., 1979 Polymorphism with migration and selection. J. Math. Biol. *8*: 73–109.

Moody, M. E., 1981 Polymorphism with selection and genotype-dependent migration. J. Math. Biol. *11*: 245–267.

Nagylaki, T., 1975 Conditions for the existence of clines. Genetics *80*: 595–615.

Nagylaki, T., 1976 Clines with variable migration. Genetics *83*: 867–886.

Nagylaki, T., 1977 *Selection in One- and Two-Locus Systems*. Springer, Berlin.

Nagylaki, T., 1978 Clines with asymmetric migration. Genetics *88*: 813–827.

Nagylaki, T. and M. E. Moody, 1980 Diffusion model for genotype-dependent migration. Proc. Natl. Acad. Sci. USA *77*: 4842–4846.

Pauwelussen, J. P., 1981 Nerve impulse propagation in a branching nerve system: a simple model. Physica D *4*: 67–88.

Pauwelussen, J. P. and L. A. Peletier, 1981 Clines in the presence of asymmetric migration. J. Math. Biol. *11*: 207–233.

Peletier, L. A., 1976 On a nonlinear diffusion equation arising in population genetics. In *Proceedings of the Fourth Conference on Ordinary and Partial Differential Equations* (ed. W. N. Everitt and B. D. Sleeman). Pp. 365–371. Springer, Berlin.

Peletier, L. A., 1978   A nonlinear eigenvalue problem occurring in population genetics. In Journées d'Analyse Non Linéaire (ed. P. Benilan and J. Robert). Pp. 170–187. Springer, Berlin.

Slatkin, M., 1973   Gene flow and selection in a cline. Genetics 75: 733–756.

Yanagida, E., 1982   Stability of stationary distributions in a space-dependent population growth process. J. Math. Biol. 15: 37–50.

DEPARTMENT OF MOLECULAR GENETICS AND CELL BIOLOGY
THE UNIVERSITY OF CHICAGO
920 EAST 58TH STREET
CHICAGO, ILLINOIS 60637

Lectures on Mathematics in the Life Sciences
Volume 20, 1989

CHAOS VERSUS NOISE-DRIVEN DYNAMICS

W.M. Schaffer and G.L. Truty

INTRODUCTION

Although much of the early work on chaos traces to ecological models (e.g., Li & Yorke, 1975; May, 1974, 1976; Guckenheimer et al., 1976; Singer, 1978), in recent years the subject has become much more the province of mathematicians and workers in the physical sciences. While this is entirely understandable--ecologists generally lack the interest and technical background for dealing with the new dynamics--it is also something of a pity. In the first place, most of the known types of chaotic behavior arise quite naturally in ecological models (Schaffer, 1987a). Moreover, the densities of natural populations, as well as the incidences of diseases that afflict their numbers, can vary a great deal (e.g., Ito, 1980). Often there is little evidence of periodicity.

To this demonstrable variability, ecologists have generally responded in one of three ways:

1. Members of the so-called "Hutchinson/MacArthur" school tend to ignore it. These workers believe quite strongly in the "Balance of Nature", which they take to be an attracting fixed point. When studying ecological models, they almost invariably set the right-hand sides of the differential equations equal to zero and solve for the equilibria. When collecting data, they focus on averages or presume that one or a few year's data are representative of some central tendency.

1980 Mathematics Subject Classification (1985 Revision). 92A17, 58F13.

2. Workers with a more statistical bent often look for contributing, or "key" factors (e.g., Varley, Gradwell, & Hassell, 1973) as determinants of population growth. The factors may either be exogenous, for example, climatic variations (Andrewarthra & Davidson, 1948), or intrinsic--generally the density of the species in question. Regarding the relative importance of "density-independent" and "density-dependent" processes, there is historic and continuing disagreement. (For review, see Tamarin, 1978.)

3. Time series analysts (e.g., Finnerty, 1980) have taken the relatively few cases in which the dynamics are more nearly periodic (for example, cycles of boreal mammals), and looked for statistical evidences of regularity.

Notice that while the first response ignores population dynamics entirely, the second and third ascribe deviations from regular behavior to noise. Thus, one is led to conclude that chance perturbations are very important. There is, however, another alternative, which is that the irregularities themselves are deterministic. Recent attempts (e.g., Olsen, 1987a,b; Schaffer, 1984, 1985, 1987a,b) to discern the fingerprints of chaos in ecological and epidemiological data sets have been motivated by this possibility. Generally speaking, these efforts presume chaos in one of its simpler guises--for example, the spiral chaos as observed in the Rössler (1976) attractor, or the nearly quasi-periodic fluctuations that follow the breakup of an attracting torus. Here, we consider the somewhat different situation in which a system spends much of its times in two more or less well-defined regions of the phase space, jumping at irregular intervals from one region to the other. Such a situation can be modelled by positing either multiple attracting fixed points plus noise or chaotic trajectories that wander in the vicinity of points that are repelling.

Before proceeding to a comparison of these alternatives, the following points are worth emphasizing:

1. Noise-driven systems with two potential wells have been suggested as models of fluctuating fishery stocks (Steele & Henderson, 1984), outbreaks of agriculturally important insect pests (Ludwig, et al., 1978), and are at the heart of Sewall Wright's (1977) shifting balance theory of evolution. More recently, Newman et al. (1985) have suggested such a model as accounting for "punctuations" in the fossil record. Notice that in all such models the motion is sustained by perturbations from

without. In the absence of noise, trajectories tend to one of the fixed points and stay there.

2. There exist chaotic systems, for example, the Lorenz (1963) equations, in which trajectories jump at irregular intervals from one "neighborhood" to another. But in this case, the fixed points are repelling. Sustained motion obtains in the absence of noise.

3. From a long-term statistical point of view, the two processes are rather similar. In particular, it is possible to construct stochastic differential equations that are natural analogs of the chaotic process (Kottalam et al., 1986).

4. In low-dimensional chaotic systems, there is often a tension between effectively stochastic behavior over long time scales and short-term determinism. For example, in systems such as the Lorenz and Rössler attractors, one can extract essentially one-dimensional maps which summarize the sequence of orbital excursions. *A priori*, one does not expect such relationships in the stochastic analogs mentioned above. This suggests a methodology for distinguishing the two kinds of behavior when the governing equations are unavailable, e.g., in the case of experimental data.

5. The foregoing possibility notwithstanding, accumulating sufficient data to make such distinctions in most ecological situations will generally be impossible. A possible exception is microbial systems where time scales are fast, and large amounts of information can be gathered.

In developing our ideas, we study the Lorenz equations, which so far as we know have never been suggested in an ecological context. The present approach is therefore purely phenomenological. On the other hand, it seems likely that ecological models with Lorenz-like dynamics could easily be constructed. Indeed, a discrete system with similar behavior has already been given for the case of single species population growth in seasonal environments (Kot & Schaffer, 1984). Here, the "basins" correspond to the population's phase relative to the seasonal cycle.

It is also hoped that the present work may be of interest inasmuch as the phenomena at issue are certainly not restricted to the Lorenz equations. In particular, we will discuss the effective discretization of continuous chaos by homoclinic bifurcations.

THE LORENZ EQUATIONS

Of all the dynamical systems exhibiting chaotic behavior, perhaps none has been so intensively studied as the system of differential equations first derived by Saltzman (1962) and later analyzed in greater detail by Lorenz (1963). The equations may be written as follows:

$$dx/dt = s (y - x) \quad dy/dt = rx - y - xz \quad dz/dt = xy - bz \quad (1)$$

Note that x, y, and z are the state variables, while s, r, and b are parameters. A good review, in fact the only place where most of what is known about this system has been gathered together, can be found in the book by Sparrow (1983; see also Guckenheimer & Holmes, 1983; and Lichtenberg & Lieberman, 1982, who discuss the derivation of (1) from fluid dynamics). Some of the essentials are as follows:

1. The Lorenz equations can be derived from the Navier-Stokes equation for Benard convection--i.e., convection in a fluid heated from below--via a three-mode truncation.

2. The resulting system of ODEs (1) is deceptively simple. Nonetheless, for values of r, the so-called "Rayleigh number", in excess of a critical value, they generate sustained, aperiodic dynamics. As reviewed by Sparrow (1983), and unlike the situation in other pur- portedly chaotic attractors such as the Rössler, there are good reasons in the case of Eqs (1) to suppose that there are whole intervals in parameter space for which the be- havior is truly chaotic.

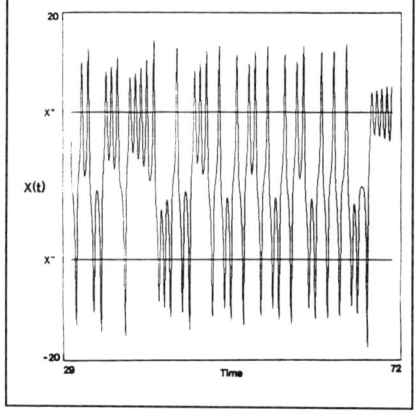

FIG. 1. Time series, X(t), for the Lorenz equations (r = 28.0; s = 10.0; b = 8/3). Horizontal lines at X^+ and X^- indicate positions of the nonstable saddle foci.

3. When plotted in the time domain (Fig. 1), trajectories exhibit diverging oscillations about one of two values. At regular intervals, the trajectory jumps from one domain to the other. Let us call the values about which oscillations diverge X^+ and X^- and the intervals between jumps RESIDENCE or FIRST PASSAGE

times. Figure 1 suggests that to first approximation, the residence times are inversely correlated with the trajectory's initial distance from X^+ or X^-. (By "initial", we refer to the distance immediately after the jump.)

4. Figure 2 shows the motion in phase space. Apparently, and this is easily demonstrated, the points X^+ and X^- are non-stable saddle foci. Between them is a third fixed point, also a saddle, which sits at the origin.

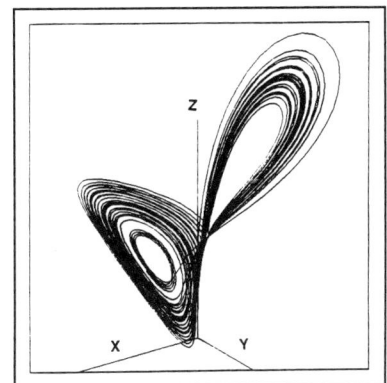

FIG. 2. A solution curve for the Lorenz equations plotted in 3-space. Parameters as in Figure 1. The axes have been scaled independently.

5. The attractor itself is organized into sheets. Locally, these are pinched together along the unstable manifold of the origin which forms the spine of what is sometimes called the "Cantor-book" (e.g., Williams, 1980). In fact, Cantor-set like structure is not visible in numerical simulations. This reflects the extremely rapid rate at which volumes are contracted under the flow. Not surprisingly, calculations of the attractor's fractal dimension (we use the term loosely) yield numbers slightly in excess of 2.0.

6. Eqs (1) exhibit both deterministic and seemingly "random" attributes. In particular:

a. The sequence and timing of orbital excursions can be summarized by discrete mappings that are effectively one dimensional (Lorenz, 1963; Yorke & Yorke, 1979; Shaw, 1981). In particular, let Z_i be the ith maximum in z(t). Plotting Z_{i+1} vs. Z_i yields the picture shown in Figure 3a. Plotting T_i, the time from one maximum to the next, vs. Z_i yields the picture shown in Figure 3b.

b. At the same time, correlations along solution curves decay rapidly (e.g., Lücke, 1979). For x(t), there are essentially no spectral peaks. This reflects the splitting effect of the interior saddle on nearby trajectories (e.g., Farmer et al., 1980).

In sum, there is the aforemention-
ed tension between short term predictability and long term "stochasticity".

7. There is an extensive
mathematical literature (Williams,
1977, 1979, 1980; Williams &
Guckenheimer, 1980) on geometric-
al models of the Lorenz attractor.
As emphasized both by Sparrow
(1983) and Guckenheimer &
Holmes (1983), the relevance of
this work to Eqs (1) remains
somewhat problematic. Specifically,
these models presume a contracting
foliation on the flow. While the
existence of such a foliation has
never been proved, it is generally
believed that much of the analysis
does, in fact, carry over.

EVOLUTION OF THE
STRANGE ATTRACTOR

The genesis and evolution
of the strange attractor, as one
varies the parameter r (with
s = 10 and b = 8/3), can be
summarized as follows:

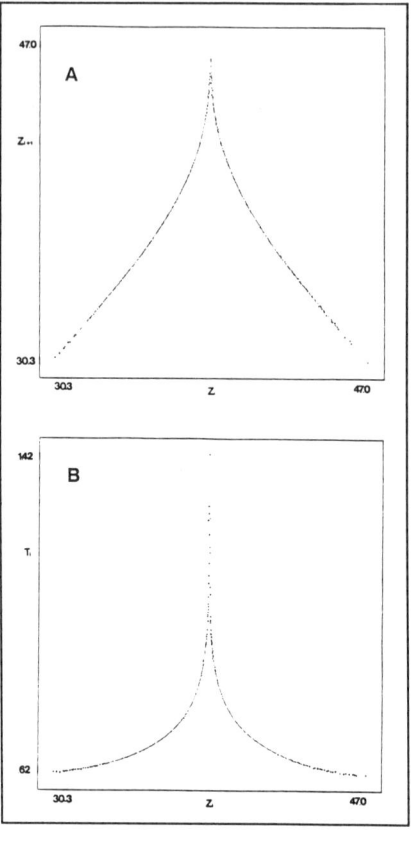

FIG. 3. Short-term determinism in the Lorenz
equations. a. Successive max-ima in z(t).
b. Time until the next maximum in z(t) vs.
magnitude of the present maximum.

1. For $1 < r < 24.74...$, the origin is nonstable, while the points
X^+ and X^- are attracting. In this parameter range, initial trajectories tend
rapidly to one of the two saddle foci.

2. Numerically computed trajectories notwithstanding, at $r =
13.926...$ the system undergoes a homoclinic bifurcation or "explosion".
This results from the existence (at this parameter value) of a homoclinic
orbit involving the origin and produces what is called a "strange invariant

set". The latter consists of a countable infinity of nonstable periodic orbits, an uncountable infinity of aperiodic orbits, and an uncountable infinity of solution curves that terminate at the origin. We emphasize that this set is not an attractor. Rather, as in the case of the Smale horseshoe, trajectories in the vicinity leave via well-defined "escape hatches".

3. At $r = 24.06...$, the escape hatches disappear. At this point, the invariant set becomes an attractor, coexisting with the two fixed points.

4. At $r = 24.74...$, the saddle foci lose stability via a sub-critical Hopf bifurcation which destroys the simplest periodic orbits created in the homoclinic explosion mentioned above. Beyond this point, nearly all initial conditions tend to the strange attractor.

5. The strange attractor which comes into being at $r = 24.06...$ undergoes continuing evolution in the sense that there is an entire (countably infinite) series of homoclinic explosions which both remove and create periodic orbits. Thus, there is not one, but many Lorenz attractors, the topology of which change continuously as the parameter is varied. In particular, for $r < 28.0$ the attractor collapses monotonically with the maximum number of revolutions about one of the saddle foci declining to 24. Beyond this point, new orbits are created which are later destroyed by cascades of period-halving bifurcations. Thus, the attractor continues to collapse, albeit in a more complicated fashion. Overall, the maximum number of revolutions about one or the other saddle focus declines as we continue to crank up the parameter.

6. Beyond $r > 313$, there is a single stable periodic orbit.

NOISE-DRIVEN OSCILLATOR ANALOG

Various authors (e.g., Zipellius & Lücke, 1981; Knobloch, 1979) have studied the statistical mechanics of the Lorenz equations and attempted to derive approximations in the form of stochastic differential systems. Most recently, Kottalam et al. (1986) have shown that the long-term statistical properties, i.e., various moments and mean first passage time, of Eqs (1) are accurately modelled by modifying Eqs (1) as follows:

Equate the derivative, dz/dt, with a Gaussian random variate and substitute for $z(t)$ in the equations for dx/dt and dy/dt. This yields:

$$dx/dt \; = \; s \, (y - x) \qquad dy/dt \; = \; rx - y - x^2y/b + u(t) \qquad (2)$$

where

$$< u(t) > \; = \; 0 \qquad < u(t) \, u(t\text{-}T) > \; = \; 2 \, D \, _d \, (T) \qquad (3)$$

This procedure is motivated by the observation (Lücke, 1976) that

$$< dz/dt > \; = \; 0 \qquad\qquad\qquad\qquad (4)$$

Equations (2) correspond to a bi-stable, noise-driven anharmonic oscillator. For suitable values of D, the system moves erratically in one of the two potential wells, jumping at irregular intervals from one basin of attraction to the other (Fig. 4). That is, the qualitative pattern is what one sees in the Lorenz equations. Interestingly, for moderately large (15,000 points) data sets, the two processes yield roughly comparable fractal dimensions (Fig. 5) using the method developed by Grassberg and Procaccia (1983).

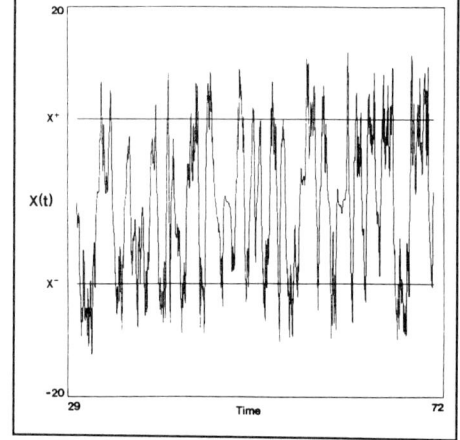

FIG. 4.   Noise-driven oscillator analog to the Lorenz equations. The time series, $X(t)$, is displayed.

FIRST PASSAGE TIMES

Here, we focus on the time between jumps. As noted above, various simple relations describing the sequence and timing of orbital excursions can be extracted from Eqs (1). Typically, these involve $z(t)$. However, in the oscillator analog, the third variable has disappeared. Hence, to compare the two systems for evidence of short-term determinism, we must look to other relationships.

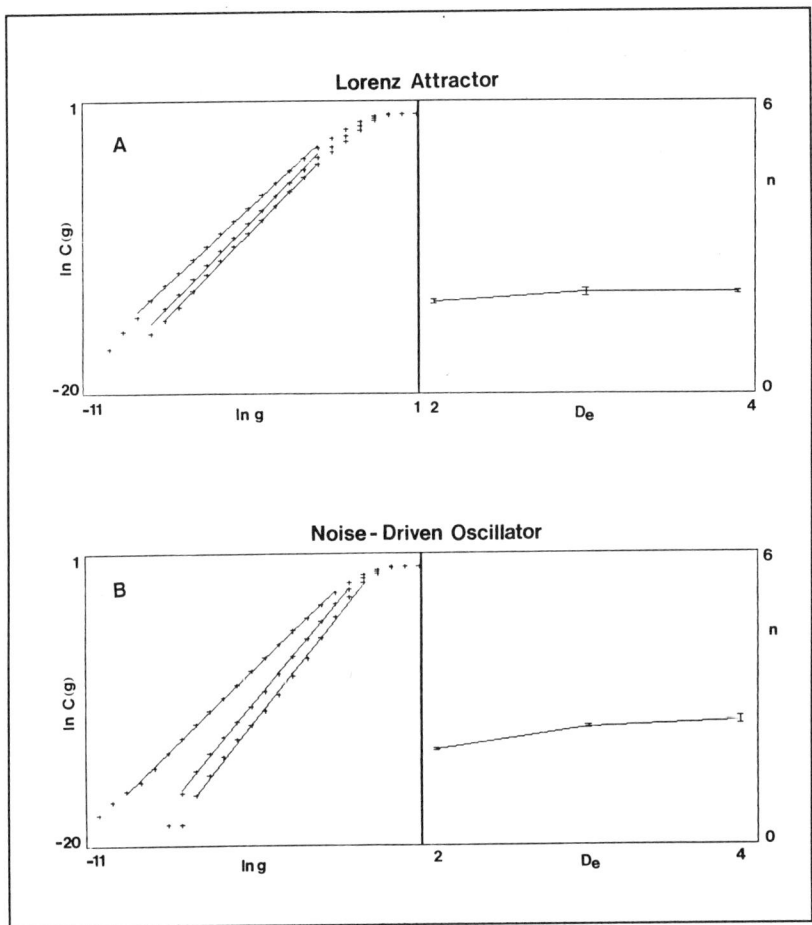

FIG. 5. Estimating the dimension of the Lorenz attractor (**top**) and a noise-driven oscillator analog (**bottom**). For each calculation, 15,000 values of x(t) were embedded in two, three, and four dimensions. **Left.** Correlation integral. **Right.** Scaling exponent, n, *vs* embedding dimension. The exponent gives a lower bound on the information dimension. For the Lorenz attractor, we calculate a dimension of about 2.05; for the noise-driven oscillator, about 2.5

Figure 6a depicts one such pattern which itself is not without interest. Here, we plot the residence times against the distance of the FIRST extremum in x(t) from X^+ or X^-. Earlier, it was noted that the residence time appears to increase as trajectories are re-injected closer to one of the saddle foci. Actually, the pattern is more complicated. Instead of a smooth function, we see what are essentially a series of discrete steps. Figure 6b gives their relative frequencies. Similar discretizations have been observed in other systems close to homoclinicity (Argoul et al., 1986).

In the case of the Lorenz equations, three points are worth noting:

1. The effective discretization of residence times was first noted by Aizawa (1978) who computed the mapping:

$$T: \quad T_i \; \text{-->} \; T_{i+1} \tag{5}$$

where T_i and T_{i+1} are successive residence times (Fig. 6c).

2. The distribution of residence times does NOT reflect the distribution of first extrema (Fig. 6d). Rather, Figure 6a can be understood by noting that each of the steps corresponds to a different number of consecutive revolutions about one of the saddle foci prior to a jump. The residence time rises abruptly each time the number of such revolutions increases.

3. As might be expected, neither the details, nor the overall distribution of residence times survives in the noise-driven oscillator analog (Fig. 7).

We can further understand the distribution of residence times as follows. For any value of r, there will be a number, i^*, which we can associate with the non-stable periodic orbits. This number is the maximum number of consecutive revolutions about one of the foci that will be on a periodic orbit. As pointed out by Sparrow (1983), the maximum number of such revolutions for an actual orbit, i.e., a trajectory based at almost any point in the phase space, cannot exceed i^*. Hence there is a natural relationship between the topology of the attractor and the number of residence times. In particular, with increasing r, there is an overall decline in i^* and thus the number of observed residence times (Figs. 8-10). This relationship is complicated by the fact that as r increases and i^* declines, the geometry of the attractor becomes more complex.

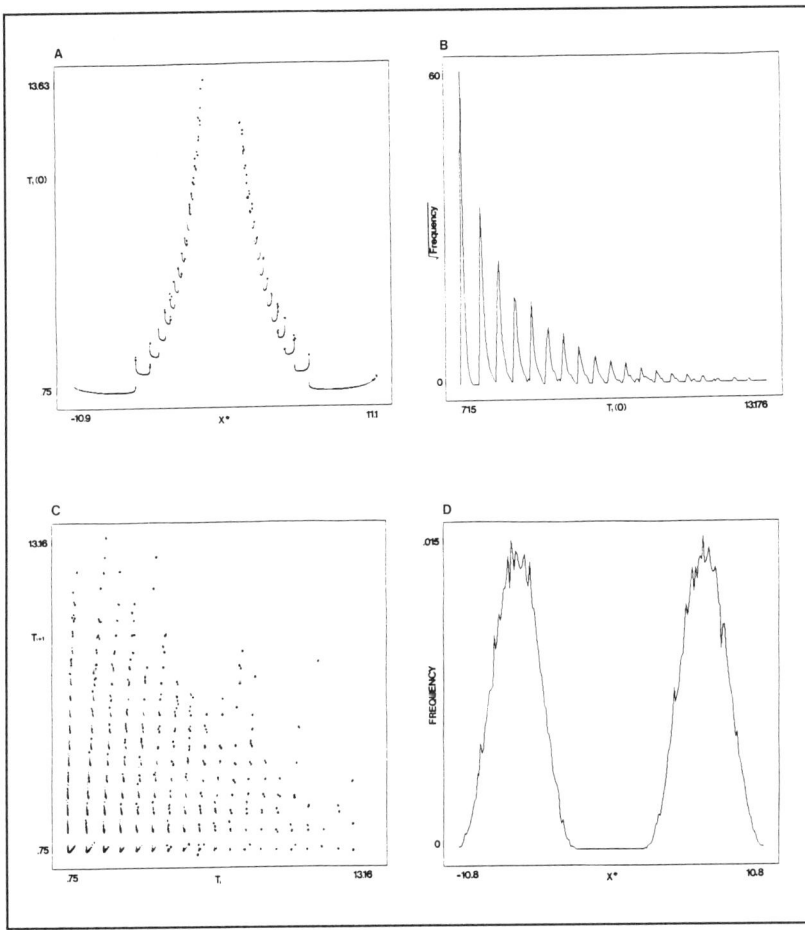

FIG. 6. Discretization of the residence time, $T_1(0)$, in the Lorenz equations. **a.** The residence times plotted against the distance, X^*, of the first extremum following a jump from the nearest saddle focus. Forty thousand jumps were accumulated. **b.** Frequency diagram. The data were divided into 200 equally-spaced bins. To increase resolution at high residence times, the square root of the frequency is displayed. **c.** Relation between successive residence times as reported by Aizawa who first noted the discretization. **d.** Frequency distribution of the first extrema following a jump. The essentially smooth distribution cannot account for the pattern observed in the $T_1(0)$.

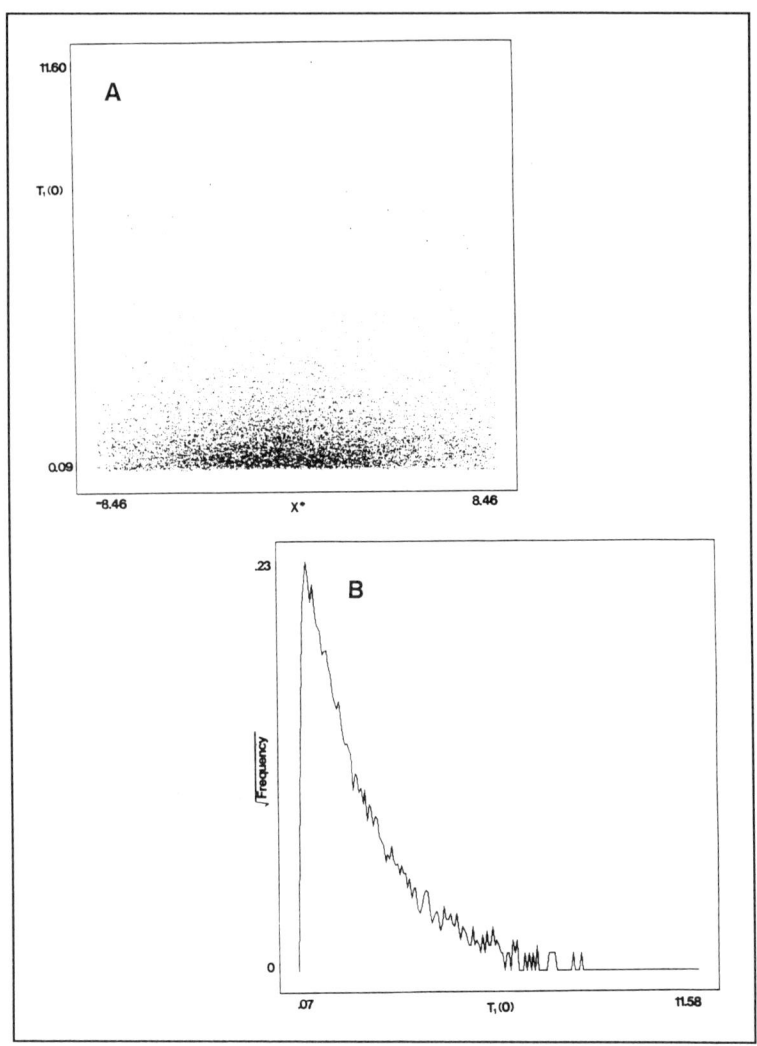

FIG. 7. Residence times for the noise-driven analog discussed by Kottalam et al. (1986). A value of $r = 78.63$ (corresponding to the observed variance in dz/dt, for $r = 28$; $s = 10$; $b = 8/3$), was used. Ten thousand flips were accumulated. Compare with Figure 5. **a.** The residence times plotted against X^*. **b.** Frequency distribution of the residence times.

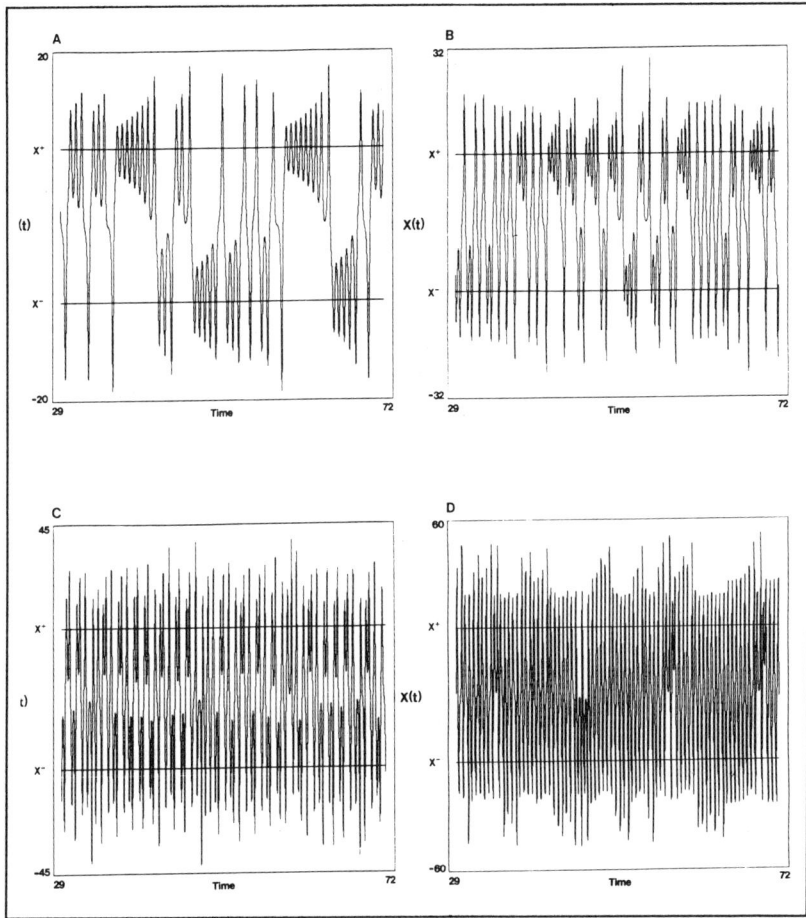

FIG. 8. Sample time series for the Lorenz equations for other parameter values. **a.** r = 30. **b.** r = 60. **c.** r = 125. **d.** r = 197.4. Note the overall decrease in the average number of revolutions about X^+ or X^- prior to a jump.

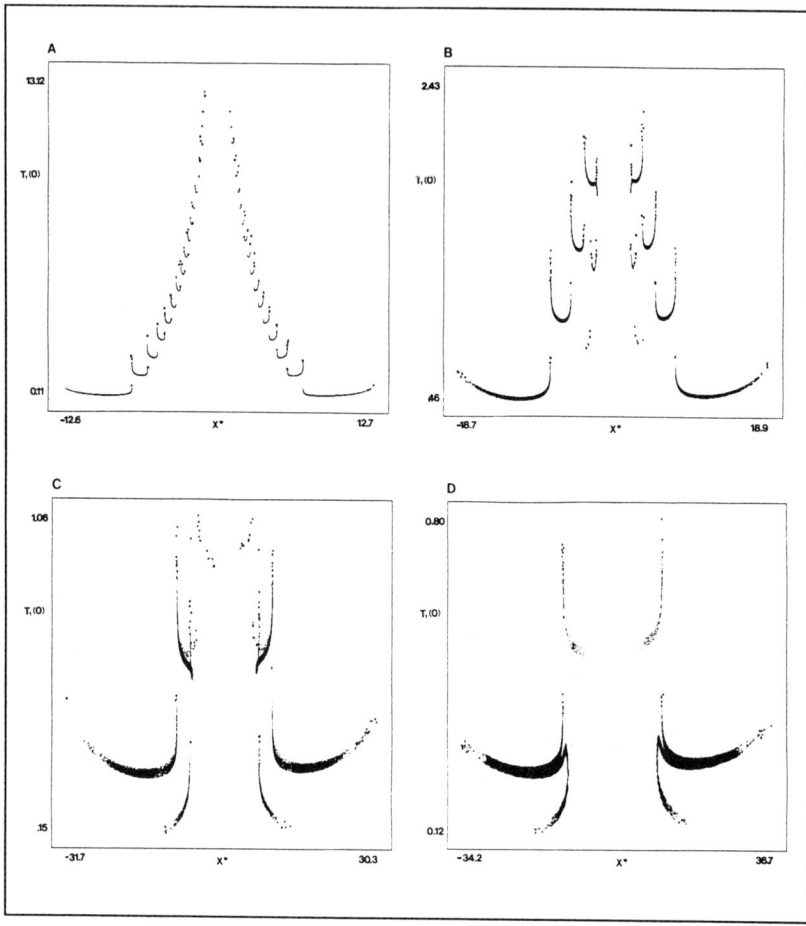

FIG. 9. Residence times plotted against X* for the r values shown in Fig. 7. In addition to a decreasing number of distinct states, one also observes (for r = 125 and 197.4) the appearance of a second set of states. These reflect increasing complexity of the attractor.

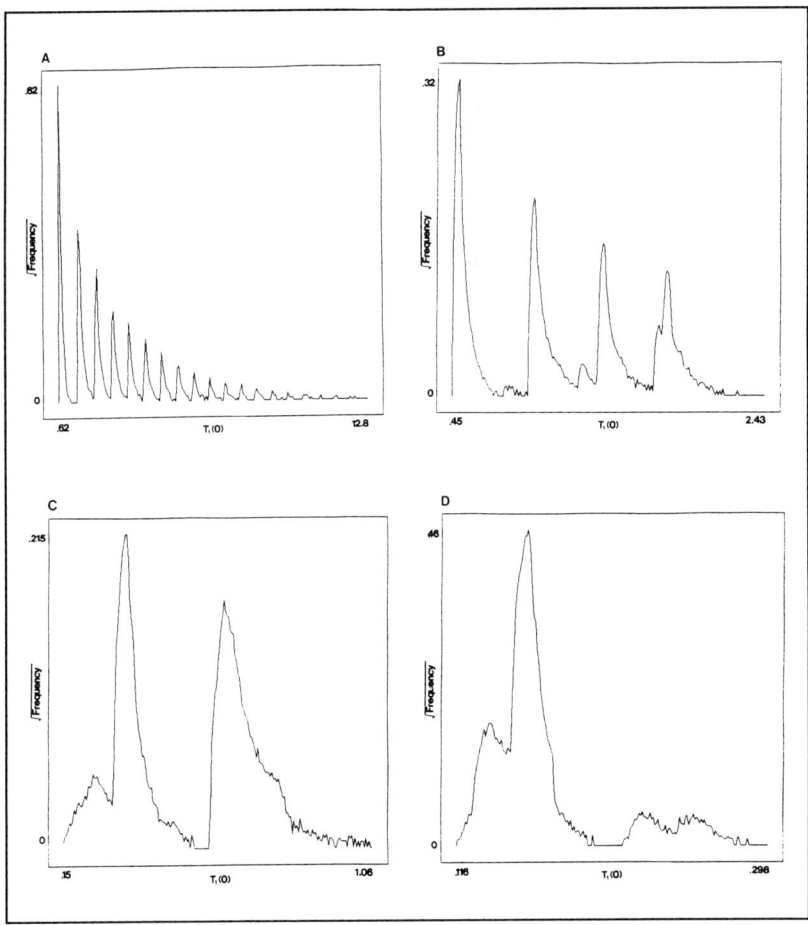

FIG. 10. Frequency distributions for the residence times shown in Fig. 7.

SUMMARY AND CONCLUSIONS

The foregoing discussion reviews some of the differences and similarities between a chaotic system and its stochastic analog. Given large amounts of data, such differences might be useful in distinguishing between the two phenomenologies. Notice that the comparison is an interesting one, being between systems that bear a natural relation, the one to the other. Beyond this, our computations illustrate the way in which homoclinicity can generate effectively discrete behavior in a continuous system.

ACKNOWLEDGEMENTS

We thank Bruce West, Mark Kot, and David Graser for discussion, and Sandra Fulmer for labelling the figures. This work was supported by grants from the National Science Foundation and the National Institutes for Health.

LITERATURE CITED

AIZAWA, Y. 1982. Global aspects of the dissipative dynamical systems. I. Progr. Theor. Phys. **68**:64-84.

ARGOUL, F., Arneodo, A., and P. Richetti. 1986. Experimental evidence for homoclinic chaos in the Belouzov-Zhabotinskii reaction. Preprint.

DAVIDSON, J. and H.G. Andrewartha. 1948. The influence of rainfall, evaporation and atmospheric temperature on fluctuations in the size of a natural population of *Thrips imaginis*. J. Anim. Ecol. **17**:200-222.

NEWMAN, C.M., Cohen, J.E., and C. Kipnis. 1985. Neo-darwinian evolution implies punctuated equilibria. Nature **315**:400-401.

FARMER, D., Crutchfield, J., Froehling, H., Packard, N., and R. Shaw. 1980. Power spectra and mixing properties of strange attractors. Ann. N.Y. Acad. Sci. **357**:453-472.

FINNERTY, J.P. 1980. The population ecology of cycles of small mammals. Yale Univ. Press. New Haven.

ITO, Y. 1980. Comparative ecology. Cambridge Univ. Press.

GRASSBERGER, P. and I. Procaccia. 1983. Measuring the strangeness of strange attractors. Physica **9D**:189-208.

GUCKENHEIMER, J. and P. Holmes. 1983. Nonlinear oscillations, dynamical systems and bifurcations of vector fields. Springer-Verlag. New York.

GUCKENHEIMER, J., Oster, G., and I. Ipaktchi. 1976. the dynamics of density dependent population models. J. Math. Biol. **4**:101-147.

GUCKENHEIMER, J. and R.F. Williams. 1980. Structural stability of the Lorenz attractor. Publ. Math. I.H.E.S. **50**:73-100.

KAPLAN, J.L. and J.A. Yorke. 1979. Preturbulence: A regime observed in a fluid flow model of Lorenz. Comm. Math. Phys. **67**:93-108.

94   W. M. SCHAFFER AND G. L. TRUTY

KNOBLOCH, E. 1979. On the statistical dynamics of the Lorenz model. J. Stat. Phys. **20**:695-709.

KOT. M. and W.M. Schaffer. 1984. The effects of seasonality on discrete models of population growth. Theor. Pop. Biol. **26**:340-360.

KOTTELAM, J., West, B.J., and K. Lindenberg. 1986. Analogy between the Lorenz strange attractor and a bistable stochastic oscillator. Preprint.

LI, T.Y. and J.A. Yorke. 1975. Period three implies chaos. Am. Math. Mon. **82**:985-992.

LICHTENBERG, A. and Lieberman, M. 1982. Regular and stochastic motion. Springer-Verlag. New York.

LORENZ, E.N. 1963. Deterministic nonperiodic flow. J. Atmos. Sci. **357**:282-291.

LUDWIG, D., Jones, D.D., and C.S. Holling. 1978. Qualitative analysis of insect outbreak systems: the spruce budworm and forest. J. Anim. Ecol. **47**:315-332.

LÜCKE, M. 1976. Statistical dynamics of the Lorenz model. J. Stat. Phys. **15**:455-475.

MAY, R.M. 1974. Biological populations with nonoverlapping generations: stable points, stable cycles and chaos. Science **186**:645-647.

-----. 1976. Simple mathematical models with very complicated dynamics. Nature **261**:459-467.

OLSEN, L.F. 1987a. Oscillations and chaos in epidemics: a non-linear dynamic study of six childhood diseases in Copenhagen, Denmark. Theor. Pop. Biol. (In press).

-----. 1987b. Low dimensional strange attractors in epidemics of childhood diseases in Copenhagen, Denmark. **In**, Degn, H., Holden, A.V., and L.F. Olsen (eds). *Chaos in biological systems.* NATO ASI Series. Life Sciences. Plenum Press. New York.

RÖSSLER, O.E. 1976. An equation for continuous chaos. Phys. Lett. **57A**:397-398.

SALTZMAN, B. 1962. Finite amplitude free convection as an initial value problem. J. Atmos. Sci. **19**:329-341.

SCHAFFER, W.M. 1984. Stretching and folding in lynx fur returns: evidence for a strange attractor in nature? Am. Nat. **124**:798-820.

-----. 1985. Can nonlinear dynamics elucidate mechanisms in ecology and epidemiology? IMA J. Math. Appl. Med. Biol. **2**:221-252.

-----. 1987a. Chaos in ecology and epidemiology. **In**, Degn, H., Holden, A.V., and L.F. Olsen (eds). *Chaos in biological systems.* NATO ASI Series. Life Sciences. Plenum Press. New York.

-----. 1987b. Perceiving order in the chaos of nature. **In**, Boyce, M. (ed). *Evolution of Life Histories: Theory and pattern from mammals.* Yale Univ. Press. New Haven.

SHAW, R. 1981. Strange attractors, chaotic behavior, and information flow. Z. f. Natüreforsch. **36a**:80-112.

SINGER, D. 1978. Stable orbits and bifurcation maps of the interval. SIAM J. Appl. Math. **35**:260-267.

SPARROW, C. 1983. The Lorenz equations: bifurcations, chaos and strange attractors. Springer-Verlag. new York.

STEELE, J.H. and E.W. Henderson. 1984. Modelling long-term fluctuations in fish stocks. Science **224**:985-987.

TAMARIN, R.H. 1978. *Population Regulation.* Dowden, Hutchinson, and Ross. Stroudsburg, PA.

VARLEY, G.C., Gradwell, G.R., and M.P. Hassell. 1973. *Insect population ecology.* Blackwell Scient. Publ. Oxford.

WILLIAMS, R.F. 1977. The structure of Lorenz attractors. In, Bernard, P. (ed). Lecture Notes in Mathematics **615**:93-113. Springer-Verlag. New York.

-----. 1979. The bifurcation space of the Lorenz attractor. Ann. N.Y. Acad. Sci. **316**:393-399.

-----. 1980. Structure of Lorenz attractors. Publ. Math. I.H.E.S. **50**:59-72.

WRIGHT, S. 1977. Evolution and the genetics of populations. Vol. III. Univ. Chicago Press. Chicago.

YORKE, J.A. and E.D. Yorke. 1979. Metastable chaos: the transition to sustained chaotic behavior in the Lorenz model. J. Stat. Phys. **21**:263-277.

ZIPPELIUS, A. and M. Lücke. 1981. The effect of external noise in the Lorenz model of the Bénard problem. J. Stat. Phys. **24**:345-358.

DEPARTMENT OF ECOLOGY
& EVOLUTIONARY BIOLOGY
UNIVERSITY OF ARIZONA
TUCSON, ARIZONA 85721

Lectures on Mathematics in the Life Sciences
Volume 20, 1989

A GENERAL MODEL OF THE ROLE OF

ENVIRONMENTAL VARIABILITY

IN COMMUNITIES OF COMPETING SPECIES

Peter L. Chesson

ABSTRACT. This paper analyzes a general model
of population dynamics for two species
competing in a fluctuating environment. The
model is a continuous state-space, discrete-
time Markov process and is studied using the
standard invasibility technique in conjunction
with a small-effects approximation. This
provides a general method of analyzing stochastic competition models and gives simple
formulae for mean low-density growth rates in
terms of mean environmental effects, mean
competitive effects, and an interaction
between environmental and competitive factors.
Hence, the manner and strength with which
environmental fluctuations affect coexistence
is exhibited in a simple interpretable form.

1. INTRODUCTION. A central concern of community

ecology is understanding the mechanisms determining

community structure (the number and kinds of species

that may live together in a particular locality). A

1980 Mathematics Subject Classification (1985
Revision). 92A17.
Work supported by NSF grant BSR-8615028.

classic essay of G. E. Hutchinson (1959) focused
attention on the phenomenon of competition as a
primary determinant of community structure. It is
generally expected on the basis of mathematical models
(see Roughgarden 1979, and May 1981) that a group of
species will not be able to coexist if competition
between species (interspecific competition), for
resources that are in short supply, is too strong. It
is known, for example, that interspecific competition
will prevent a stable equilibrium coexistence when the
per-capita growth rates of different species in a
community are functions of the same factor (Armstrong
and McGehee 1980).   This single factor could, for
example, be a common food source for which the organisms compete, or in plants it could be light, moisture
or a nutrient.

   The extent to which interspecific competition may
be a factor in community structure has been the subject of intense debate recently (e.g. Den Boer 1986,
Abrams 1986, Giller 1986, Roughgarden 1986). An
important consideration is the fact that most populations in nature are subject to varying environmental conditions which may upset the predictions of
equilibrium models based on competition alone.

   Following the early work of May and MacArthur
(1972), there have been a number of attempts to dev-
elop a theory of community structure accounting for
the effects of environmental variability (Woodin and
Yorke 1975, Armstrong and McGehee 1976, Levins 1979,
Turelli 1981, Chesson and Warner 1981, Chesson 1983,
1984, Ellner 1984, Shmida and Ellner 1985, Abrams
1985). While this work has produced some striking
general conclusions, there has been some confusion

about the circumstances under which environmental
variability should promote coexistence or should do
the opposite: promote competitive exclusion.
Moreover, most theoretical findings have been qualitative. The quantitative detail useful for checking
the predictions of models in nature has been lacking.

It is the purpose of this article to continue the
development of a general model put forward in Chesson
(1987) that appears to provide a simple way of understanding the effects of environmental variability on
community structure. Moreover, this model yields
useful quantitative predictions. In Chesson (1987) it
was necessary to assume that all of the species in the
system would be biologically very similar. In this
paper this assumption is removed, allowing consideration of the full range of possible ways in which
species may differ from one another.

2. THE MODEL. The basic assumptions of the model are
that the growth rate of a population can be expressed
as a function of environmental and competitive
factors. It is assumed that the major environmental
effects can be summarized by a single population
parameter such as a denisty-independent birth rate,
survival rate or germination rate. This environmentally-dependent parameter will be denoted by $E_i(t)$,
indicating that it is specific to species i and varies
with time t. It will always be defined so that larger
values reflect more favorable conditions. Thus, E
will never be a mortality rate but could be a survival
rate.

The effects of competition are also assumed
representable by a single number $C_i(t)$ summarizing the

combined effects of intraspecific and interspecific
competition on species i. For instance, competition
might be measured by a comparison of the number of
individuals competing for a limited amount of a
critical resource and the supply of that resource.
The precise form of $C_i(t)$ will depend on the
circumstances, but in general it will be assumed to be
a function of the population densities of the species
in the system, $X_1(t)$, ..., $X_n(t)$, and their environmentally-dependent parameters. Thus

$$(1) \qquad C_i = c_i(E_1, X_1, E_2, X_2, \ldots, E_n, X_n),$$

where c_i is some function of 2n variables. Time, t,
is suppressed for notational simplicity.

   Environmentally-dependent parameters can be
expected to affect the amount of competition occurring
through their effects on the abundances of competing
forms. For instance, if the environmentally-dependent
parameter is a germination rate, it will affect the
abundance of seedlings competing as they grow, and
thus will strongly influence the intensity of
competition. The function c_i is assumed to be
monotonic in its arguments, and in particular it is
assumed to increase as a function of population
density. In general, if $X_j = 0$, i.e., species j is
absent from the system, c_i is assumed to be constant
in the argument E_j.

   Population dynamics of the species in such a
community will be represented in discrete-time form as

$$(2) \qquad X_i(t+1) = G_i(E_i(t), C_i(t))X_i(t),$$

where the function G_i combines the environmentally-
dependent parameter and the competitive parameter to

obtain the finite rate of increase of the population
from time t to time t+1. Generally, G_i will be an
increasing function of E, and a decreasing function of
C.

Analysis of models of populations in stochastic
environments is facilitated by expressing population
change on a log scale where we obtain

$$(3) \qquad \ln X_i(t+1) - \ln X_i(t) = g_i(E_i, C_i),$$

with $g_i = \ln G_i$. Note that g_i is analogous to r in
demography, and we shall refer to it simply as the
growth rate. This growth rate is important because by
summing it over any period of time we obtain the
change in log population size for that period. Thus
the average of the growth rate over any given period
indicates whether the population increases or
decreases over that period.

The important properties of the model are
determined by the way in which E and C interact in
their specification of the growth rate. In the case
of no interaction (the additive case), the growth rate
can be expressed in the form

$$(4) \qquad g_i(E_i, C_i) = A_i(E_i) + B_i(C_i),$$

for some functions A_i and B_i. This additive form
applies whenever the quantity

$$(5) \qquad \gamma_i \overset{\text{def}}{=} \frac{\partial^2 g_i}{\partial E_i \partial C_i}$$

is equal to zero. When this quantity is negative, the
growth rate is called subadditive; when positive the
growth rate is called superadditive. The intuitive
meanings of these different possibilities are

discussed in Chesson (1987). In brief, when γ is negative (the subadditive case), poorer environmental conditions make the growth rate less sensitive to competition: g_i does not decline so quickly as a function of C. Essentially, the growth rate is buffered against simultaneous poor conditions such as a poor environment and high competition. As shown in Chesson (1987), simple yet common features of biology can produce these buffered growth rates.

When (5) is negative, the growth rate is called superadditive. In this case the growth rate is more sensitive to competition in a poorer environment: a poor environment in effect amplifies competition.

To expose the effects of the interaction between environment and competition, the parameters of the model can be transformed so that the model takes the following standard form:

$$(6) \qquad g_i(E_i, C_i) = E_i - C_i + \Gamma_i(E_i, C_i)$$

(Chesson 1987). To see how to obtain this form, we designate the original parameters and growth rate function using the subscript o, and we then choose some particular value E_i^* of the original environmentally-dependent parameter E_{io}, and a corresponding value C_i^* of C_{io} such that $g_{io}(E_i^*, C_i^*) = 0$. These define conditions that would constitute equilibrium in a constant environment. The new parameters and growth rate function are expressed in terms of the originals as

$$(7) \qquad E_i = g_{io}(E_{io}, C_i^*), \quad C_i = -g_{io}(E_i^*, C_{io}),$$

and

(8) $g_i(E_i, C_i) = g_{io}(E_{io}, C_{io})$.

This standard form minimizes the differences between
different models. In particular, it ensures that
environmental and competitive factors have similar
meanings in terms of their effects on the growth rate
in all models.   Indeed, near the values (E_i, C_i) =
$(0, 0)$ (corresponding to (E_i^*, C_i^*) in the old
variables), the environmentally-dependent parameter
and competitive parameter respectively represent the
translation of environment and competition into the
growth rate of the population. This standard form
allows a unified treatment of the general class of
models given by equation (2). Unless otherwise specified, we assume below that the model has been transformed to this standard form.

3. ANALYSIS OF THE MODEL. To analyze the model, we
restrict attention to the two-species situation and
make the assumption that the environment process
$\{(E_1(t), E_2(t)), t = 0, 1, \ldots\}$ is an i.i.d. process,
i.e., different times are independent and the bi-
variate distribution of $(E_1(t), E_2(t))$ is independent
of t. It follows that the community process
$\{(X_1(t), X_2(t)), t = 0, 1, \ldots\}$ is a homogeneous Markov
process and can be analyzed using the standard invasibility analysis reviewed in Chesson (1987). This
involves seeing if a species at low density (an
"invader") can increase in the presence of the other
species (the "resident"). The resident is minimally
affected by the invader while the invader remains at
low density. The Markov process describing the
dynamics of the resident, with the invader density set

equal to 0, can usually be expected to converge on a
stationary distribution (Ellner 1984). We let c_i^j be a
random variable representing the competitive parameter
of species i when j is at its stationary distribution
and species i is at zero density. The law of large
numbers implies that this invading species, i, will
have a trend to increase from low density if it has a
positive value of

$$(9) \qquad \Delta_i \overset{\text{def}}{=} \quad Eg_i(E_i, \ c_i^j).$$

This quantity is the mean low-density growth rate, or
"boundary" growth rate of species i. A positive
boundary growth rate is usually interpreted as imply-
ing that the species persists in the system. Thus, if
the boundary growth rates of both species are
positive, the species are interpreted as coexisting.
Indeed, for a broad class of models there is strong
mathematical justification for this assumption
(Chesson and Ellner 1988). Moreover, there is good
reason to expect the case $\Delta_i > 0$, $\Delta_j < 0$, to mean that
species i persists in the system while species j
converges to extinction.

The final case, defined by Δ_1, $\Delta_2 < 0$, can be
expected to imply that one of the species will become
extinct while the other persists, with no certain
predictability of which species does which. These
assertions have only been proved for special classes
of models within this general framework (Chesson and
Ellner 1988), but are generally assumed to apply
broadly. For the purposes of discussion, we shall
assume that we are in situations where these assertions are valid. However, none of the mathematics

below depends on this assumption.

When a species is present alone, the assumptions
of the model imply that there will generally be a
unique equilibrium density, X^*, for any given constant
value of the environmentally-dependent parameter.
This equilibrium density is the solution of the
equation $C_i = 0$.

For the analysis, we introduce some additional
assumptions.

(a)   The fluctuations in $E_i(t)$ are concentrated in
        a finite range about 0, with length propor-
        tional to a small parameter σ.

(b)   When the system is simplied to contain just a
        single species, a unique stationary distri-
        butions exists, and the corresponding equil-
        ibria, X_i^*, for constant-environment cases, are
        asymptotically exponentially stable.

(c)   The mean values (EE_i) of the environmentally-
        dependent parameters are of order σ^2 $(O(\sigma^2))$.

(d)   The function c_i satisfies $c_i(0, 0, 0, X_j^*) = O(\sigma^2)$.

(e)   The random variable $c_i(0, 0, E_j, X_j)$ can be
        written as a function of $c_j(E_j, X_j, 0, 0)$.

(f)   The regression of E_i on E_j is linear to order
        σ^2, i.e. $E[E_i|E_j] = EE_i + b_e(E_j - EE_j) + O(\sigma^2)$, where b_e is a constant.

Assumption (a) is simply the small fluctuation
assumption. Assumption (b) insists on regular
behavior for the single-species systems. Assumptions
(c) and (d) imply that constant factors affecting the
growth rates of the species, as represented by the
means of the environmentally-dependent parameters and
the competitive differences between species, have

similar strengths to the environmental variance, which
is proportional to σ^2. This means that these constant
factors are of smaller magnitude than the fluctuations
in the environmentally-dependent parameter, which are
of order σ, by assumption (a).

Assumption (e) implies that the intraspecific and
interspecific competitive effects caused by a species
are determined by a common underlying function of its
environmentally-dependent parameter and population
size. For example if E_j is a birth rate and competition occurs among offspring, then $E_j X_j$ will be the
number of offspring of species j, which will determine
both the intraspecific and interspecific competitive
effects of species j.

Assumption (f) simply expresses the natural ex-
pectation of approximately linear relationships when
fluctuations are restricted to small ranges. It is
consistent with the other assumptions of the model.

Standard differentiability assumptions are introduced in the appendix where it is shown that the
stationary distribution of the single-species system
gives fluctuations in the competitive factors C_i that
are concentrated in a finite range of order σ about 0.
For the single-species system it follows that g_i has
the second order Taylor approximation

$$(10) \qquad g_i(E_i, C_i) \approx E_i - C_i + \gamma_i E_i C_i,$$

where $\gamma_i = \partial^2 g_i / \partial E_i \partial C_i$, evaluated at $(0, 0)$, and "$\approx$"
means that the difference between the LHS and RHS is
$o(\sigma^2)$. The appendix now shows that equation (10)
continues to apply when species i is an invader (at
zero density) and species j is a resident at its
stationary distribution. These results also imply

that $E[c_i^j | c_j^j]$ is linear to $0(\sigma^2)$. The corresponding linear regression coefficient will be denoted by b_c.

To perform the invasibility analysis, we note that the expected value of (10) must be zero for a resident at its stationary distribution, because the expected value of $\ln X_i(t)$ cannot change once the stationary distribution has been reached. Defining χ_{ij} as the covariance between E_i and c_i^j and making use of the fact that $EE_j c_j^j = EE_j E c_j^j + \chi_{jj} = 0(\sigma^2) + \chi_{jj}$, we obtain

$$(11) \qquad E c_j^j \approx EE_j + \gamma_j \chi_{jj}.$$

Taking expected values in (10) under the assumption that species i is at zero density, and species j is a resident, we obtain

$$\Lambda_i \approx EE_i - E c_i^j + \gamma_i EE_i c_i^j$$

$$(12) \qquad \approx EE_i - E c_i^j + \gamma_i \chi_{ij}.$$

Now

$$\chi_{ij} \approx E\{E[E_i c_i^j | E_j]\} \approx Eb_e E_j c_i^j \approx b_e b_c \chi_{jj},$$

and so

$$(13) \qquad \Lambda_i \approx EE_i - E c_i^j + \gamma_i b_e b_c \chi_{jj}.$$

Note that in this equation, $E c_i^j$ will likely be dependent on environmental variability, (c.f. equation [11] for the case of intraspecific competition), and so equation (13) may have environmental variability occurring twice but in ways that are in opposition. To overcome this problem, we substitute in (13) the quantity $\Delta C = E c_i^j - E c_j^j$, representing the difference

between average interspecific and average intraspecific competition. While this quantity may still
depend on environmental variability, its dependence is
likely to be less than the absolute average values of
the these competitive parameters. Making this sub-
stitution in (13), using equation (11), defining $\theta = \gamma_i/\gamma_j$, and $\Delta E = EE_i - EE_j$, we obtain

$$(14) \qquad \Delta_i \approx \Delta E - \Delta C - \gamma_j(1 - \theta b_e b_c)\chi_{jj}.$$

Thus the boundary growth rate is expressed in components that can be regarded as relatively independent.
The first term represents mean environmental effects,
the second term represents the mean excess of interspecific competition over intraspecific competition,
and the last term represents the effects of the interaction between environment and competition.

Although expression (14) seems fairly clean, it is
not always the best representation of the different
effects because ΔC can be nonzero when different
species respond to exactly the same competitive factor
but have different magnitudes of response. In particular, ΔC can be nonzero when different species have
growth rates that differ only by a positive
proportionality constant, meaning that when one
species increases, the other must also. In such cases
the proper value of the difference in competitive
effects should be 0. To make this the standard by
which competitive differences between species are
compared, one can define

$$(15) \qquad \Delta C = EC_i^j - qEC_j^j,$$

where q is some constant. A suitable choice for q
will often be b_c, because it measures the magnitude of

response of species i to competition compared with the
response of species j to the same factor. Having done
this, it is appropriate to redefine the environmental
differences ΔE as $EE_i - qEE_j$. Equation (14) becomes

(16) $\Delta_i \approx \Delta E - \Delta C - \gamma_j(q - \theta b_e b_c)\chi_{jj}.$

Note however, that equation (14) is a special case of
equation (16).

This equation shows that the boundary growth rate
is simply the sum of three parts: (a) the mean environmental advantage that species i has over the
other species, (b) the mean excess of interspecific
competition over intraspecific competition, and (c) an
effect due to the interaction between environmental
variability and competition. The covariance χ_{jj} is a
reflection of variation in the environmentally-
dependent parameter of the competitor species, not a
direct reflection of the variability in the environmentally-dependent parameter of the species in
question. As shown below, in the case of an i.i.d
environment, this covariance is simply proportional to
the variance in the competitor's environmentally-
dependent parameter.

The term in parentheses contains a mixture of
terms dealing with the asymmetry of the two species
and the correlations between their environmentally-
dependent parameters. To gain an appreciation of this
term, it is helpful to consider a special case in
which the species are similar enough that they have
the same g_i, have q = 1 (reflecting the same magnitude
of response to competitive conditions) and have equal
variances for their environmentally-dependent parameters. In this case the term in parentheses reduces

to 1 - r, where r is the correlation between the
environmentally-dependent parameters of the two
species. Thus the term in parentheses will be positive unless the species respond in exactly the same
way to the environment, in which case it will be 0.

In many cases the covariance χ_{jj} will be positive,
for example when the environmental factor does not
directly affect competitive processes (Chesson 1987).
For simplicity, we will restrict discussion to this
case. Assuming that asymmetry is not too strong (or
that the correlation between the environmentally-
dependent parameters is low), the term in parentheses
will be positive too. This means that a negative
interaction between environmental and competitive
effects will lead to a positive addition to the boundary growth rate in the presence of environmental
variability. This positive addition can make up for
environmental and competitive disadvantages that a
species may experience, and therefore can permit it to
coexist with a superior competitor. In this setting
environmental variability will promote species diversity.

In the opposite case, where the interaction term
is positive, environmental variability will lead to a
reduction in boundary growth rates and coexistence
will be demoted. Indeed, species that could coexist
in the absence of environmental variability may now
both have negative boundary growth rates. This would
mean that when either one of the species fluctuates to
low density it will have a negative average growth
rate, indicating an expected decline toward extinction. If the amount of environmental variability is
of intermediate magnitude and there are asymmetries in

the competitive abilities of the two species, one species could have a negative boundary growth rate while that of the other remains positive. In this situation, environmental variability contributes to the competitive elimination of the inferior species, while the other species persists.

Scenarios other than those painted above are possible from this equation. The covariance could be negative in some settings, which would reverse the above conclusions. Alternatively, it might be negative for one species but positive for the other. Then environmental variability would have opposite effects on the two species, unless the interactions were also of opposite sign.

While each of the three terms constituting equation (16) can be regarded as conceptually distinct from the others, it will not always be possible to vary them independently. However, while we remain within a general framework where the g_i and C_i are not assumed to take particular forms, the different quantities in equation (16) can indeed be considered independent. In this case, the parameter space is infinite dimensional and there are no constraining relationships among the finite set of parameters appearing in equation (16). In specific applications, however, the model will often be defined in terms a finite number of underlying parameters. Indeed, sometimes the number of such parameters will be small. Whenever, the number of parameters defining the model is less than the number appearing in expression (16), there will necessarily be constraints among the latter. Indeed, it may not be possible to vary the three terms in this equation independently. See section 5 below.

4. THE COVARIANCE. To calculate the covariance, χ_{jj}, we simply express the resident's parameter in terms of its first-order Taylor expansion in E_j and X_j about $(0, X_j^*)$:

(17) $C_j^j = c_{j1}E_j + c_{j2}(X_j - X_j^*) + O(\sigma^2)$.

It follows that

(18) $\chi_{jj} \approx c_{j1}VE_j$.

This formula depends critically on the assumption that the environment process has no autocorrelation. This means that $X_j(t)$ and $E_j(t)$ are uncorrelated, and so the second term in (17) does not affect (18). In the autocorrelated case the formula is

(19) $\chi_{jj} \approx c_{j1}VE_j + c_{j2}C(E_j, X_j)$,

where the covariance $C(E_j, X_j)$ is calculated using the joint stationary distribution of E_j and X_j. In this autocorrelated case, expression (16) is not always correct and requires a slight modification. Expression (16) remains true, however, in a number of useful cases of autocorrelation, in particular, if the environment processes of the two species are independent of each other.

5. BACK TRANSFORMATION. Often it will be desired to know how the results of the model appear in terms of the original parameters, rather than just the standard parameters. In particular, this will facilitate the investigation of the constraints among the quantities in equation (16). In this section we derive formulae

that permit the results above to be expressed in terms of the original parameters.

To distinguish the standard parameters and the original parameters, the original parameters and all quantities associated with them will be given the subscript o for "original." We define

$$(20) \qquad \alpha_i^{(n)} = \left. \frac{\partial^n}{\partial E_{io}^n} g_{io}(E_{io}, C_{io}^*) \right|_{E_{io} = E_{io}^*}.$$

A quantity $\beta_i^{(n)}$ is defined similarly as the nth derivative of $-g_{io}$ in its second argument. In case of the first derivative, the superscript (n) will sometimes be dropped.

We can now obtain an approximation for the standard parameters in terms of the orginal parameters by using Taylor series expansion of g_{io}. This yields

$$(21) \qquad E_i \approx \alpha_i^{(1)}(E_{io} - E_{io}^*) + \frac{1}{2}\alpha_i^{(2)}(E_{io} - E_{io}^*)^2,$$

with a corresponding expression for C_i. It is easily seen that

$$(22) \qquad \gamma_i = \gamma_{io}/\alpha_i\beta_i,$$

$$(23) \qquad \chi_{jj} \approx \alpha_j\beta_j\chi_{jjo}$$

$$(24) \qquad b_e = (\alpha_i/\alpha_j)b_{eo}, \quad b_c = (\beta_i/\beta_j)b_{co}$$

and

$$(25) \qquad \theta = \theta_o(\alpha_j\beta_j/\alpha_i\beta_i).$$

Although it will generally be possible to express q in terms of the original parameters, in general the quantity "q_o" is not meaningful and is best considered undefined.

With these definitions, the components of equation
(16) can be expressed in terms of the original parameters.   In doing this we use the notation "$\hat{\ }$" to indicate the deviation from a *-value, e.g. $\hat{E} = E - E*$.
We obtain

$$(26) \qquad \Delta E \approx \alpha_i^{(1)} E\hat{E}_{io} - q\alpha_j^{(1)} E\hat{E}_{jo}$$

$$+ \tfrac{1}{2}(\alpha_i^{(2)} VE_{io} - q\alpha_j^{(2)} VE_{jo}),$$

and

$$(27) \qquad \Delta C \approx \beta_i^{(1)} E\hat{C}_{io}^j - q\beta_j^{(1)} E\hat{C}_{jo}^j$$

$$+ \tfrac{1}{2}(\beta_i^{(2)} VC_{io}^j - q\beta_j^{(2)} VC_{jo}^j).$$

Finally

$$(28) \qquad \gamma_j(q - \theta b_e b_c)\chi_{jj} \approx \gamma_{jo}(q - \theta_o b_{eo} b_{co})\chi_{jjo}.$$

The first thing to note is that the interaction
term, which is the main focus of this work, takes the
same form in terms of the original parameters as it
does in terms of the standard parameters.   Thus choice
of parameterization does not lead to ambiguity in the
interpretation of this term.   The other quantities, ΔE
and ΔC, take forms which will be different depending
on the particular parameterization.   For example, the
variances of the parameters can appear here in terms
of the original parameters, but are not present when
standard parameters are used.   Thus these terms can be
sensitive to the variances of the original parameters
to varying degrees and in varying ways depending on
the particular situation.   It is therefore important
to consider what different parameterizations will mean

before one attempts to interpret the results.

Example: The Lottery Model. For purposes of illustration we consider the application of these results to an especially well-studied model, the lottery model (Chesson and Warner 1981, Chesson 1982, Hatfield and Chesson 1988). The lottery model is a model of competition for space. We shall consider it in the form in which the environmentally-dependent parameter is the birth rate (which we take to include early juvenile mortality). The model can then be written as

(29) $G_i(E_i, C_i) = 1 - \delta_i + E_i/C_i$

where δ_i is the adult death, and is assumed constant here. Competition occurs only among juveniles as they seek sites for settlement among the places given up by adult death in the previous time period. Thus, the formula for C_i is

(30) $C_i = (E_1X_1 + C_2X_2)/(\delta_1X_1 + \delta_2X_2),$

which is equal to the amount of space opened up by adult death divided by the number of juveniles competing for this available space.

While this form of the model reveals its assumptions most clearly, simpler and more interpretable results are abtained if the birth rate is expressed on the natural log scale. In particular, this scale more clearly elucidates the constraints among the parameters that analysis of a specific model is intended to reveal. All that is changed by this transformation is the way the results are expressed, not the results themselves. A further transformation simplifies the analysis with no effect on the form of the results of the model: the natural log of (24) multiplied by δ_i is used as the competitive factor. With these changes the model takes the form

$$(31) \qquad G_{io} = 1 - \delta_i + \delta_i e^{E_{io} - C_{io}}.$$

Simple calculations show that

$$(32) \qquad \alpha_i^{(1)} = \beta_i^{(1)} = \delta_i$$

and

$$(33) \qquad \alpha_i^{(2)} = -\beta_i^{(2)} = -\gamma_{io} = \delta_i(1 - \delta_i).$$

Moreover,

$$(34) \qquad c_{jo}^j = E_{jo} \quad \text{and} \quad c_{io}^j = E_{jo} + \ln(\delta_i/\delta_j),$$

which implies the identical relationship among the means, equality of the variances and also that $\chi_{jjo} = VE_{jo}$, and $b_{co} = 1$.

The final quantity that we need is q, which we take here to be b_c. To get it we note that the standardized competition parameter takes the form

$$(35) \qquad c_j^j = -\ln(1 - \delta_j + \delta_j e^{E_j^* - E_{jo}})$$

and

$$(36) \qquad c_i^j = -\ln(1 - \delta_i + \delta_i e^{E_i^* - E_{jo}} - \ln \delta_i/\delta_j).$$

Using the fact that

$$(37) \qquad E_i^* - E_j^* = \ln \delta_i/\delta_j$$

we deduce that $q = \delta_i/\delta_j$. Substituting in (23-24) and rearranging, we find that

(38) $\Delta E / \delta_i = E\hat{E}_{io} - E\hat{E}_{jo} + \frac{1}{2}[(1-\delta_i)VE_{io} - (1-\delta_j)VE_{jo}]$

(39) $\Delta C / \delta_i = \frac{1}{2}[(1-\delta_j) - (1-\delta_i)]VE_{jo}$,

(40) $\gamma_{jo}(q-\theta b_{eo} b_{co})\chi_{jjo} = \delta_i[(1-\delta_i)b_{eo} - (1-\delta_j)]VE_{jo}$.

This shows that the three components of the boundary
growth rate involve common factors, and so varying any
one of these common factors can lead to changes in
each of the components of the boundary growth rate.
For instance, increasing VE_{jo} by itself will increase
ΔE, will increase or decrease ΔC depending on the sign
of $(1-\delta_j) - (1-\delta_i)$, and increase or decrease the
interaction term (37) depending on the sign of
$(1-\delta_i)b_{eo} - (1-\delta_j)$.

A significant question is whether the three terms
(38-40) can vary independently as the environment is
manipulated. The answer to this is clearly yes if the
means, variances and covariances of the environmentally-dependent parameters can be varied arbitrarily, but the answer is generally no if only some
of these quantities can be varied. For instance, if
the correlation between the environmentally-dependent
parameters of the two-species is held fixed, then the
possible values of the interaction term are constr-
ained by the value of the competition term. In par-
ticular, if the correlation is 0, and $\delta_i \leftrightarrow \delta_j$, then
the value of ΔC determines the value of the inter-
action term. The fact that the competition term is
not 0 implies that a variable environment introduces
differences between interspecific and intraspecific
competition that are absent when the environment is
constant.

When expressions (38-40) are combined, one finds

that even though environmental variability possibly
pushes different components of the boundary growth
rate in different directions, overall the effect is
very simple. We obtain

(41) $\Delta_i / \delta_i = E\hat{E}_{io} - E\hat{E}_{jo} + \frac{1}{2}(1-\delta_i)V(E_{io} - E_{jo})$.

Thus we see that any disadvantage that species i may
have to species j, in terms of the mean of its
environmentally-dependent parameter, can be overcome
by variance in the difference between the environmentally-dependent parameters. It is interesting to
note that (41) agrees closely with exact numerical
results for the lottery model given in Chesson and
Warner (1981) for a broad range of parameter values.

6. DISCUSSION.   The most important feature of the
results obtained here is the ability to obtain quanti-
titive statements on the conditions for coexistence
without making overly restrictive assumptions on the
relationships of the organisms to each other. The
fact that the different species can have different
competitive factors permits the simultaneous consideration several mechanisms of coexistence. The main
focus of this work has been on how the interaction
between environment and competition can alter average
growth rates in the presence of environmental variability. This present model allows other factors to be
operating through the difference between the mean
interspecific and intraspecific effects. For example,
the mean interspecific effects might tend to be
smaller than the mean intraspecific effects, and this
would be favorable to coexistence with or without
environmental variability. This case could arise

through differences in resource requirements of the
different species ("resource partitioning") or could
be a reflection of frequency-dependent predation
(Roughgarden and Feldman 1974), even though the biological interpretation of these conditions is quite
different.

The standard parameterization is responsible for
the unity of the results even though the underlying
models can differ arbitrarily and the environmentally-
dependent parameters may represent different things in
different circumstances. This standard parameterization permits the boundary growth rate (the mean
growth rate at low density) to be expressed as a sum
of three meaningful components, which clearly show how
the different factors that affect a species appear in
this boundary growth rate. It is important to note,
however, that within the constraints imposed by the
details of a specific system, it may not be possible
to vary all three components independently. Moreover,
depending on the scale on which environmental variability is measured, all of them may change with
changes in the amount of environmental variability.
Only the critical term involving the interaction
between environment competition is independent of the
scale of measurement. The form of the results
obtained here suggests a standard scale of measurement
removing the ambiguity inherent in models involving
environmental variability when there is no rationale
for measuring variability on any particular scale.

7. APPENDIX: SINGLE-SPECIES RESULTS. We need to show
that the single-species iterations confine C to a
finite range of order σ. As it causes no confusion,

we drop the subscript i throughout. We assume that $g(E, C)$ and $C(E, X)$ ($= c(E, X, 0, 0)$) are jointly first differentiable in their arguments with first derivatives bounded in finite intervals. We define $Z(t) = \ln X(t)$ and $f(E, Z) = g(E, C(E, e^Z))$. For simplicity of notation we assume that $E^* = C^* = 0$, and $|E| \leqslant \sigma$.

Asymptotic exponential stability of the single-species case in a constant environment means that there is a symmetric neighborhood (z_1, z_2) of z^*, and a positive number $\rho < 1$ such that

(a1) $|z - z^* + f(0, z)| \leqslant \rho|z - z^*|$.

Setting $z' = z + f(E, z)$ and substituting in (a1) we get

(a2) $|z' - z^*| \leqslant \rho|z - z^*| + |f(E, z) - f(0, z)|$

(a3) $\leqslant \rho|z - z^*| + \delta$,

where δ is the supremum over $|E| \leqslant \sigma$ and $z \in (z_1, z_2)$.

The differentiability assumptions mean that $\delta = O(\sigma)$ and so for σ sufficiently small we have $z' \in (z_1, z_2)$. This means that

(a4) $Z(t) \in (z_1, z_2)$

whenever $Z(0) \in (z_1, z_2)$. Moreover,

(a5) $\limsup |Z(t) - z^*| < \dfrac{\delta}{1 - \rho}$.

As a consequence of (a5) and the bounded differentiability of C, it follows that

(a6) $\limsup |C(E(t), X(t))| \leqslant K\sigma$

for some constant K. Assumption (d) implies the identical result for the case where $C(E(t), X(t))$ is interspecific competition experienced by an invader with $E(t)$ and $X(t)$ referring to the resident.

The formulae for back transformation require that

$E\hat{E}_{io}$, $E\hat{C}^j_{jo}$, and $E\hat{C}^j_{io}$ are each $O(\sigma^2)$. This can be
deduced easily by using the corresponding properties
of the standardized parameters. That $EC^j_j = O(\sigma^2)$
follows from equation (11); and assumption (d) now
implies the same for EC^j_i.

Acknowledgements   I am grateful for comments on the
manuscript from Stephen Ellner, Nancy Huntly, and
Andrew Taylor.

BIBLIOGRAPHY

1.   Abrams, P. 1984. Variability in resource consump-
tion rates and the coexistence of competing species.
Theoret. Pop. Biol. 25: 106-124.

2.   Abrams, P. 1986. Letter to the editor, Trends in
Ecology and Evolution 1, 131-132.

3.   Armstrong, R. A., and R. McGehee. 1976. Coexistence of species competing for shared resources.
Theoret. Pop. Biol. 9: 317-328.

4.   Armstrong, R. A., and R. McGehee. 1980. Competitive exclusion. Amer. Natur. 115: 151-170.

5.   Chesson, P. L. 1982. The stabilizing effect of a
random environment. J. Math. Biol. 15: 1-36.

6.   Chesson, P. L. 1983. Coexistence of competitors in
a stochastic environment: the storage effect. In H. I.
Freedman, and C. Strobeck, eds., Population Biology,
Lecture Notes in Biomathematics No. 52, pp. 188-198.

7.   Chesson, P. L. 1984. The storage effect in stochastic population models. In S. A. Levin and T. G.
Hallam, eds., Mathematical Ecology: Trieste Proceedings, Lecture Notes in Biomathematics No. 54, pp.
76-89.

8.   Chesson, P. L. 1987. Interactions between environment and competition: how fluctuations mediate coexistence and competitive exclusion. Lecture Notes in
Biomathematics, in press.

9.   Chesson, P. L., Ellner, S. P. 1988. Invasibility
and stochastic boundedness in monotonic competition
models. J. Math. Biol., in press.

10. Chesson, P. L., and R. R. Warner. 1981. Environmental variability promotes coexistence in lottery competitive systems. Amer. Natur. 117: 923-943.

11. Ellner, S. P. 1984. Asymptotic behavior of some difference equation population models. J. Math. Biol. 19: 169-200.

12. DenBoer, P. J. 1986. The present status of the competitive exclusion principle. Trends in Ecology and Evolution 1, 25-28.

13. Giller, P. 1986. Letter to the editor, Trends in Ecology and Evolution 1, 132.

14. Hatfield, J., Chesson, P. L. 1988. Diffusion approximation and stationary distribution for the lottery competition model. Manuscript.

15. Hutchinson, G. E. 1959. Homage to Santa Rosalia or why are there so many kinds of animals? Amer. Natur. 93, 145-159.

16. Levins, R. 1979. Coexistence in a variable environment. Amer. Natur. 114: 765-783.

17. May, R. M., and R. H. MacArthur. 1972. Niche overlap as a function of environmental variability. Proc. Natl. Acad. Sci. USA 69: 1109-1113.

18. May, R. M., ed. 1981. Theoretical Ecology, Sinauer Associates, Sunderland.

19. Roughgarden, J. 1979. Introduction to the Theory of Population Genetics and Evolutionary Ecology.

20. Roughgarden, J. 1986. Letter to the editor, Trends in Ecology and Evolution 1, 132.

21. Roughgarden, J., Feldman, M. W. 1974. Species packing and predation pressure. Ecology 56, 489-492.

22. Shmida, A., and S. P. Ellner. 1985. Coexistence of plant species with similar niches. Vegetatio 58, 29-55.

23. Turelli, M. 1981. Niche overlap and invasion of competitors in random environments I. Models without demographic stochasticity. Theoret. Pop. Biol. 20: 1-56.

24. Woodin, S. A., and J. A. Yorke. 1975. Disturbance, fluctuating rates of resource recruitment, and increased diversity. In S. A. Levin ed., Ecosystem Analysis and Prediction, Proceedings SIAM–SIMS Conference, Alta, Utah, 1974, pp. 38–41.

DEPARTMENT OF ZOOLOGY
THE OHIO STATE UNIVERSITY
1735 NEIL AVENUE
COLUMBUS, OHIO 43210

ABCDEFGHIJ-89